The Blame Machine: Why Human Error Causes Accidents

R.B. Whittingham

ELSEVIER
BUTTERWORTH
HEINEMANN

AMSTERDAM   BOSTON   HEIDELBERG   LONDON   NEW YORK   OXFORD
PARIS   SAN DIEGO   SAN FRANCISCO   SINGAPORE   SYDNEY   TOKYO

Elsevier Butterworth-Heinemann
Linacre House, Jordan Hill, Oxford OX2 8DP
200 Wheeler Road, Burlington, MA 01803

First published 2004

British Library Cataloguing in Publication Data
Whittingham, Robert B.
  The blame machine : why human error causes accidents
  1. Accidents   2. Errors   3. Accidents - Case studies
  I. Title
  363.1

Library of Congress Cataloguing in Publication Data
Whittingham, Robert B.
  The blame machine : why human error causes accidents / Robert B. Whittingham.
    p. cm.
  Includes bibliographical references and index.
  ISBN 0-7506-5510-0 (alk. paper)
    1. Accidents. 2. Accidents–Prevention. 3. Human engineering. I. Title.

HV675.W48 2003
363.1'065–dc22

               2003049531

ISBN 0 7506 5510 0

> For information on all Elsevier Butterworth-Heinemann publications
> visit our website at www.bh.com

Typeset by Charon Tec Pvt. Ltd, Chennai, India
Printed and bound in Great Britain

Contents

Preface

The village where I live is situated close to a north–south trunk road. The main route out of the village connects to the slip road for the trunk road going northbound, the layout of the junction being shown in Figure P1. However, to reach the northbound lane of the trunk road, a driver needs to turn right and proceed southbound along the slip road which then turns through 180 degrees to go north. Conversely, to go south along the trunk road, the driver must turn north along the slip road, turn right to cross the bridge over the trunk road and then turn right again to proceed down the south-bound slip road on the other side of the trunk road. For the first few months that I lived in the village, if I wanted to go north I instinctively turned north along the slip road, only to find I was heading for the southbound carriageway! Realizing my error, I had

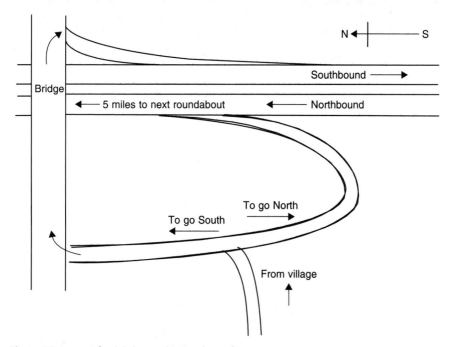

Figure P1 Layout for joining major trunk road.

to retrace my route. The same happened in reverse. If I wanted to go south and I turned south on the slip road out of the village I ended up going north down the trunk road. Then I had to drive 5 miles to the next roundabout to reverse direction. Eventually, I got the hang of it, but even now, I need to think carefully before I make the crucial first turn. My sense of north and south directs my way finding due to my inadequate mental map of the road layout.

Some human errors appear to be completely random. For instance, when carrying out a routine task such as making a cup of tea in the morning, not quite awake and slightly bleary-eyed I sometimes pour the milk into the teapot instead of the cup. There is absolutely no explanation for this except perhaps that my mind was elsewhere or I was distracted. However, most errors, such as the error of turning the wrong way down the slip road are not random but are caused by the system, in this case the road layout. Errors that are not random, but are caused by the system are referred to as 'system induced' or 'systemic' errors. The book deals principally with systemic errors because it may be possible to correct the system to reduce the chance of an error. In this case, maybe a north–south signpost for traffic leaving the village would help the motorist unfamiliar with the road layout.

This is a rather trivial example of the way human error can be precipitated by an external cause. The error is trivial for two reasons. Firstly, the consequences of the error are slight, a mere inconvenience to the driver. Secondly, the error is recoverable since the driver only needs to go to the end of the road to turn round and begin his journey again. Unfortunately, there are many more examples of human errors where the consequences are not trivial and/or the error is not recoverable. The book will use case studies of non-trivial errors to demonstrate how and why they have occurred and what could have been done to reduce their likelihood of occurrence or the severity of their consequences. However, the objective of the book is not simply to reproduce case studies of systemic human errors that have led to serious accidents, interesting as this may be. Rather, it is intended to identify common features in the accidents and the way they are investigated, so that lessons may be learned to prevent similar accidents in the future.

Anyone who has studied human error becomes aware that apart from the system faults that cause the error, there is another aspect which needs to be considered. This is the aspect of blame. In the immediate aftermath of a serious accident there is a natural tendency, especially by the media, quickly to suggest a cause and maybe attribute blame. Quite often the words 'it is believed the accident was a result of human error' are heard. There is an air of finality about this statement, it being implicit that no more can be said or done. The inevitable has happened. Hopefully the accident investigation which is held later does not accept human error as inevitable but goes on to reveal the underlying reasons why the error occurred.

But there are many exceptions to this approach. Organizations, and sometimes whole industries, become unwilling to look too closely at the system faults which caused the error. Instead the attention is focused on the individual who made the error and blame is brought into the equation. Examples of this will be given in the book. There are many reasons why it happens. Firstly, to blame an individual for what happened may provide the simplest explanation and preclude a lengthy and expensive investigation. Unfortunately there are cases where the individual who made the error did not survive the accident and is therefore unable to put up any defence against the attribution of blame. Examples of this are also included in the book. Another common reason why an organization might wish to blame an individual, is that it diverts attention from system faults that may be inconvenient or expensive to correct. Whatever the reason, the 'The Blame Machine' swings into action and the consequence is always the same. Further accidents of a similar nature will occur because the underlying causes have not been corrected. This approach has only short-term cost benefits. Apart from the human cost of future accidents in terms of loss of life, injury and trauma, the long-term cost to the organization in loss of production, customers and reputation will far exceed the cost of correcting a faulty system.

Another aspect of blame is the conflict between the desire to arrive at the truth and the desire of individuals to protect themselves from blame, deserved or undeserved. The desire to avoid blame is a natural human tendency, particularly in the emotive atmosphere of the immediate aftermath. However, in the more measured atmosphere of the accident investigation and inquiry process the legalistic implications of accepting blame come to the fore. The quasi-legal, inquisitorial nature of the UK judicial inquiry system does not lend itself to finding the truth when reluctant witnesses, frightened that they may be blamed for the accident are hesitant to come forward or tell the whole story. Conversely witnesses may have evidence which they would like to put forward but do not get the chance. As in any Court of Law, witnesses are restricted to answering only the questions that are put to them. Legal counsel acting on behalf of the clients who are paying them may not ask questions which jeopardise their client's case and this could include questions which would put their client's systems or technology under scrutiny. I know examples of witnesses who had additional evidence or views about the cause of an accident, and who had revealed this to an investigator, yet they were not allowed to put this information before the court. If it is not put before the court it will not appear in the findings of the inquiry or in the inquiry proceedings and important aspects of an accident may go unrecorded. The book is not intended to re-invent the public inquiry system, but merely to point out some of its shortcomings through the many accident case studies included.

It is human to make mistakes and in any task, no matter how simple, errors will occur. The book will show that the frequency at which they occur will depend on the nature of the task, the systems associated with the task (whether they be physical or organizational systems) and the influence of the environment in which the task is carried out. It has been estimated that about 80 per cent of all accidents are due to human error in some form. It is the reason why the study of human error is crucial to preventing accidents. It is crucial both to the prevention of major accidents with multiple fatalities which make the headlines with sickening regularity, as well as to the host of minor accidents leading to injury and disability, which rarely make the headlines, but still cause untold human suffering. The names of the many major accidents due to human error are imprinted indelibly in the public memory. More recently, the rail accidents at Southall, Paddington, Hatfield and Potters Bar come to mind, the first two accidents caused by train driver error, the others from faults in the organization of the privatized railway. Further back in time, and perhaps now fading from the public memory, are the names of Zeebrugge (1987, the capsize of the Herald of Free Enterprise ferry – 192 deaths due to business misconduct), Piper Alpha (fire and explosion on oil platform – 167 deaths due to maintenance errors), Kings Cross (1987, underground fire – 31 deaths due to lack of adequate emergency procedures, poor maintenance and unsafe design), Clapham Junction (triple train collision – 35 deaths due to maintenance error) and Flixborough (1974, chemical plant explosion – 28 deaths due to management error). The list could be extended indefinitely and many examples will be used as case studies in the book to demonstrate why human error caused the accidents and how they might have been prevented.

The book adopts a pragmatic approach to the subject and is intended for non-specialists. In this respect it differs from some other textbooks dealing with human error that concentrate on the complex, psychological causes. It focuses instead upon the system faults that can make errors more likely. Part I of the book classifies the different types of errors which can occur and suggests ways of estimating their frequency so that their incidence may be minimized by recognizing error potential where it exists. Part II of the book comprises a wide range of historical case studies from various industries in order to show how systemic errors have in the past led to catastrophic accidents. The use of these case studies also enables the more technical subject matter in Part I of the book to be better understood. In order to change the future, the lessons of history need to be brought to bear on the present. This is why the book draws so heavily on what has happened in the past in the hope that some of the terrible accidents referred to in the case studies may never be repeated.

Acknowledgements

The author gratefully acknowledges the help of many people in the preparation of this book, either by way of encouragement when the task seemed formidable or by providing technical advice, knowledge or ideas when the subject seemed difficult. In particular I would like to thank the following.

David J. Smith (Author and Consultant) for his guidance in the formative stages when I was putting together the structure of the book.

Paul Hollywell (Ergonomist and Engineer) for always finding time in his busy life to respond to me when I wanted to test out an idea on him.

The late Ralph King (Author, Scientist and Investigator) from whom I learned about perseverance and never taking 'no' for an answer.

My wife Patricia for her unwavering support and unstinting help in reading and checking manuscripts (as well as thinking of a title for the book).

The many human factors experts and colleagues with whom I have worked at various times and places over the years and from whom I have learned so much (they know who they are). And to those same people I apologise if they think I have made this complex subject too simple or cut corners to get a point across to the less experienced reader.

I

Understanding human error

To err is human

1.1.1 Introduction

Everyone makes mistakes. Human errors are a part of our everyday experience. A human error could therefore be defined quite simply as 'someone making a mistake'. The reality is much more complex and before this book can proceed much further, it is necessary to produce some clear definition of human error and the way it is manifested. Many have tried and some have succeeded in defining human error. Some examples from various sources follow and are listed under the name of the author. In studying these definitions, it should be noted that each author has a distinct purpose in mind when formulating his definition and that the definition will be useful within that context. The objective here is to produce a final definition which will be suitable within the context of this book.

1.1.2 Swain and Guttman 1983

An error is an out of tolerance action, where the limits of tolerable performance are defined by the system.

Swain and Guttman 1983

This is an interesting definition because it allows the system response to determine whether an error has occurred. Thus a human error is a deviation from normal or expected performance, the deviation being defined by the consequence. The consequence is some measurable characteristic of the system whose tolerable limits have been exceeded, rather than the human action that contains the error. However, after the error has been made, the human action within which the error occurred can be examined to determine the cause of the deviation. Also useful here is the concept of an out of tolerance action, indicating that there are limitations to human performance which can be accepted without a human error having necessarily occurred.

1.1.3 Reason 1990

A generic term to encompass all those occasions in which a planned sequence of mental or physical activities fails to achieve its intended outcome, and when these failures cannot be attributed to some chance agency.

Reason 1990

Again, this definition focuses on the outcome or consequence of the action rather than on the action itself in order to determine if an error has occurred. In this definition it is recognized that the desired end result may follow a pre-planned sequence of human actions, which has to take place successfully before the result is achieved. Any one or more of the actions in the sequence may contain an error that causes the intended outcome not to be achieved. This closely reflects the reality of many industrial situations. The definition is also interesting in that it excludes random or chance events from the category of human error. This is discussed in more detail below.

1.1.4 Hollnagel 1993

An erroneous action can be defined as an action which fails to produce the expected result and/or which produces an unwanted consequence.

Hollnagel 1993

Hollnagel prefers to use the term 'erroneous action' rather than 'human error'. The problem, according to Hollnagel, is that human error can be understood in different ways. Firstly, it can refer to the cause of an event, so that after an accident occurs, it is often reported that it was due to human error. Human error can also be a failure of the cognitive (or thinking) processes that went into planning an action or sequence of actions, a failure in execution of the action or a failure to carry out the action at all. Erroneous action defines what happened without saying anything about why it happened.

1.1.5 Meister 1966

A failure of a common sequence of psychological functions that are basic to human behaviour: stimulus, organism and response. When any element of the chain is broken, a perfect execution cannot be achieved due to failure of perceived stimulus, inability to discriminate among various stimuli, misinterpretation of meaning of stimuli, not knowing what response to make to a particular stimulus, physical inability to make the required response and responding out of sequence.

Meister 1966

This quite detailed definition perceives human actions as comprising three elements:

- Stimulus – the perception by the senses of external cues which carry the information that an action should be carried out.
- Organism – the way these stimuli are interpreted, the formulation of an appropriate action and the planning of how that action should be carried out.
- Response – the execution of the planned actions.

As with Reason's definition, this emphasizes the reality that no single human action stands alone, but is part of a sequential process and that human error must be understood in the context of this. This principle will become abundantly clear as human error is examined in the light of accident case studies. When the events that precede a human error are found to have an influence on the probability of the error occurring, the error is referred to as a human dependent failure. In addition, although a human error may represent a deviation from an intended action, not every error necessarily leads to a consequence because of the possibility of error recovery. In fact many errors are recoverable, if they were not, then the world would be a much more chaotic place than it actually is. Error recovery is an extremely important aspect of the study of human error and will be dealt with in more detail later in this book, as will human error dependency.

1.1.6 Characterizing an error

1.1.6.1 *Intention to achieve a desired result*
A common element in all the above definitions is that for a human error to occur within an action, the action must be accompanied by an intention to achieve a desired result or outcome. This eliminates spontaneous and involuntary actions (having no prior conscious thought or intent and including the random errors which are discussed in more detail below) from the category of human errors to be considered in this book. To fully understand spontaneous and involuntary errors it is necessary to draw upon expertise in the fields of psychology, physiology and neurology, disciplines which are beyond the scope of this book which offers, as far as possible, a pragmatic and engineering approach to human error. Readers interested in delving further into these topics can refer to more specialist volumes.

1.1.6.2 *Deciding if an error has occurred*
One way of deciding whether or not an error has occurred is to focus on the actual outcome and compare this with the intended outcome. Then, it could be said, if the intended outcome was not achieved within certain limits of tolerability, an error has occurred. Such a definition would, however, exclude the important class of *recovered errors* mentioned above. If an error occurs but its effects are nullified by some subsequent recovery action, it would be incorrect to say that this was not a human error and need not be investigated. It is possible that on a subsequent occasion, the same error might occur and the recovery action not take place or be too late or ineffective, in which case the actual outcome would differ from the intended outcome. It must be

true to say that the initiating error was always an error even if the final outcome was as intended, if it was necessary to carry out an intervening action successfully for the intended outcome to be achieved. The subject of error recovery is considered in detail in a later chapter.

1.1.6.3 *The significance of an error*

An important principle to be established in the study of human error is that *the significance or severity of a human error is measured in terms of its consequence*. A human error is of little or no interest apart from its consequence. In one sense, a human error that has no consequence is not an error assuming that recovery has not taken place as discussed above. There is nothing to register the occurrence of an error with no consequence except for the perception of the person making the error assuming the person was aware of an error being made. At the same time, an error does not have to be special or unique to cause an accident of immense proportions. The error itself may be completely trivial, the most insignificant slip made in a moment of absentmindedness or an off-the-cuff management decision. The seriousness of the error or the decision depends entirely on the consequences. This principle makes the study of human error not only important but also challenging. If any trivial human error is potentially capable of causing such disproportionate consequences, then how can the significant error that will cause a major accident be identified from the millions of human errors which could possibly occur. Significant error identification will be discussed later in this book.

1.1.6.4 *Intention*

An extremely important aspect of 'what characterizes an error' is the degree of intention involved when an 'out of tolerance action' is committed. It is important because later in the book a distinction is made between errors and violations. The violation of a rule is considered as a separate category to that of an error. A violation of a rule is always an intentional action, carried out in full knowledge that a rule is being disobeyed, but not necessarily in full knowledge of the consequences. If a violation is not intentional then it is an error. It is important to make the distinction because the method of analysis of violations (proposed later in this book) differs from the normal methods of human error analysis which are also described. The difficulty is that some classes of violations verge on error and are quite difficult to differentiate. The best method of making the distinction is to assess whether the action was intentional or not. For the purposes of this book, a human error is by definition always considered to be unintentional.

1.1.6.5 *A final definition*

A final definition of human error which suits the purposes of this book yet which takes into account the above characteristics and some of the other definitions given above, is proposed as follows:

> *A human error is an unintended failure of a purposeful action, either singly or as part of a planned sequence of actions, to achieve an intended outcome within set limits of tolerability pertaining to either the action or the outcome.*

With this definition, a human error occurs if:

(a) there was no intention to commit an error when carrying out the action,
(b) the action was purposeful,
(c) the intended outcome of the action was not achieved within set limits of tolerability.

With this definition in place, it is now possible to examine how human error can be classified using a number of error types and taxonomies. First of all, however, it is necessary to make an important distinction between random errors, which are not considered in this book, and systemic errors which are considered.

1.2 Random and systemic errors

1.2.1 Introduction

Although random errors are not the main subject of this book, it is necessary to examine them briefly here in order to be able to distinguish them from systemic errors. The characteristics of a random error (adopted for the purposes of this book) are that it is unintentional, unpredictable and does not have a systemic cause (an external factor which caused the error or made it more likely). The source of a random error will be found within the mental process and will therefore be difficult to identify with any certainty and even more difficult to correct with any prediction of success. This is discussed in more detail below.

Although random errors are by definition unpredictable they are not necessarily improbable. It is indeed fortunate that most human errors are not truly random events that occur unpredictably in isolation from any external point of reference. If this were the case, then the identification and reduction of human error might well be made impossible and there would be little purpose in writing this book. Fortunately most human errors have underlying systemic causes that can be identified, studied and at least partly addressed in order to make the errors less likely. It is only this possibility that makes a whole range of dangerous human activities acceptable.

1.2.2 Error causation

Two types of human error causation can be postulated and are referred to simply as:

- internal causes leading to *endogenous* error,
- external causes leading to *exogenous* error.

Endogenous errors have an internal cause such as a failure within the cognitive (or thinking and reasoning) processes. Some writers refer to these internal causes as 'psychological mechanisms'. In order to explain the occurrence of endogenous errors, it would be necessary to draw upon insights from the psychological, physiological or neurological sciences.

By contrast, exogenous errors have an external cause or are related to a context within which a human activity is carried out such as aspects of the task environment which might make an error more likely. However, even exogenous errors require internal cognitive processes to be involved. The difference is, that in an exogenous error, some feature of the external environment has also played a part in causing the error. This could happen for instance in the situation where the person responding to a stimulus is presented with confusing or conflicting information. The mental interpretation and processing of this information is then made more difficult, the planned response is not appropriate and results in an exogenous error. Conversely when an endogenous error occurs, there is at least no evidence of an external cause although it is difficult to show this with certainty.

Although the distinction between endogenous and exogenous errors may seem rather artificial, it is nevertheless an important concept for understanding the nature of human error. It is important because exogenous errors are theoretically capable of being reduced in frequency through changes to the external environment, while endogenous errors are not.

1.2.3 Human performance

In practice it is a matter of judgement whether an error is exogenous or endogenous since there will never be complete information about the cause of an error. One way of making the judgment is to assess whether the external environment or stimulus to action seems conducive to reasonable performance, given the capabilities of the person undertaking the task. If it is, then the error may well be endogenous in nature. However, if it is judged that a reasonable person, having the requisite skills would be unable to undertake the task successfully, then the error is almost certainly exogenous in nature. Human performance is therefore a function of the balance between the capability of the person carrying out the task and the demands of the task. The achievement of good performance consists in obtaining the right balance as illustrated in Figure 1.1.

Although it may not be possible to predict the occurrence of random errors, it may still be possible to estimate their frequency. Many random errors seem to occur at the extremes of human variability. As an example, we might imagine a well-motivated person, supported by a well-designed system, working in a comfortable (but not too comfortable) environment. The person carries out a fairly simple but well practised routine, one which demands a reasonable but unstressed level of attention and which

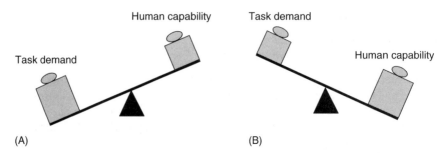

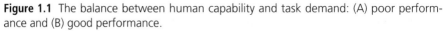

Figure 1.1 The balance between human capability and task demand: (A) poor performance and (B) good performance.

retains concentration and interest. Most of the time the task will be carried out successfully. However, there will be rare occasions when that person may well commit an inadvertent and inexplicable error. This can almost certainly be classed as an endogenous or random error.

1.2.4 Estimating human error probability

Basic probability theory and the methods of allocating actual probability values to human errors is discussed in more detail in later chapters of this book. The reason for estimating human error probability is that it provides a benchmark for measuring the benefits of improvements made to the systems that support human performance. This is particularly the case in safety critical situations such as operating a nuclear power station, driving a train or in air traffic control. In general, quantification of human error is feasible in the case of exogenous errors, but less so in the case of endogenous or random errors.

One approach to quantification of human error, which will be discussed later in the book, is to assume an average or mean probability of error for a particular type of task such as selecting a rotary control from a group of similar controls. The actual probability of error in a given situation can then be assessed by examining human capability (which may or may not be average) versus the demands of the task, as discussed in Section 1.2.3. The demands of the task may be assessed by looking for instance at how the group of rotary controls are laid out and how clearly they are labelled. If they are not laid out logically or they are not clearly labelled, then the demands of the task will be much greater and so will the error probability. Conversely, if the person making the selection is not sufficiently trained or experienced, then a higher probability of error may be expected. Although the demands of the task may be acceptable, the scales may still become unevenly balanced if human capability is insufficient.

It is a constant theme of this book that the causes of exogenous errors are deficiencies in the systems in place to support the person carrying out the task, or indeed the absence of such systems. Thus exogenous errors resulting from the failure or inadequacy of systems will be referred to as system induced or *systemic error.* Conversely, in accordance with the pragmatic nature of this book, random errors are not generally considered since their probability is indeterminate and they are less susceptible to being corrected.

1.2.5 The balance between random and systemic errors

The term *system* is defined in its widest sense to mean a physical system such as a work environment, a human/machine interface, such as a control panel for machinery, a computer display or a set of written procedures, or even a non-physical system such as a company organizational structure, a working culture or management style. Whatever, the systems that are in place to support the human they must be to a sufficiently high standard if good human performance is to be achieved. However, it needs to be understood that even with the best possible systems in place, human error will never be entirely eliminated. There will always be a residual or base level of unpredictable or random errors. Above this residual level there will be a level of systemic errors depending

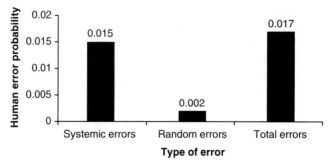

Figure 1.2 Comparison of systemic and random (residual) error before system improvements.

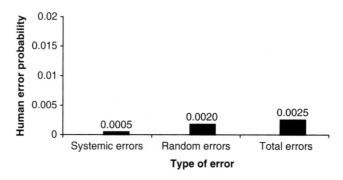

Figure 1.3 Comparison of systemic and random (residual) errors after system improvements.

on the adequacy of the systems which support (or fail to support) human performance. This principle is shown in Figures 1.2 and 1.3.

For a particular, but undefined task, Figure 1.2 shows that the total probability of an error is made up of two components, systemic errors and random errors. The probability of a systemic error is shown as 0.015. This represents a probability or chance of a systemic error of about 1 in 70 (the measurement of human reliability in terms of error probability is discussed in more detail in a later chapter). Thus an error will occur on average about once in every 70 occasions that the task is carried out. The probability of a random error is shown as 0.002 or a chance of an error of about 1 in 500. Since the causes of random error are, by definition, indeterminate and not dependent on the task demands, there is a high-level of uncertainty in the assessment of such a probability value.

Figure 1.2 shows that the total error probability is 0.017, equivalent to about a 1 in 60 chance of an error occurring every time the task is carried out. This is obtained by using probability theory and sums the probabilities of the systemic and random errors (see Chapter 5, Section 5.2 for a more detailed discussion of probability theory). Since the systemic error probability is 1 in 70, the contribution of the random element to the total probability of 1 in 60 is quite small (about 12 per cent). These probability values are not based on any particular type of task but are assigned merely to demonstrate the principles involved. It would be difficult in an actual situation to determine the relative probabilities of each type of error. However, it can be stated that an error probability of about 1 in 70 for the systemic error would be considered rather high for a critical or important task which is carried out at a high frequency.

Since the systemic error probability is theoretically due to external causes, in particular the systems associated with performance of the task, it is reasonable to suppose that some improvements to the systems might be considered. If significant improvements could be made, then the probability of the systemic error component might be reduced as shown in Figure 1.3.

Referring to Figure 1.3, the systemic error probability, as a result of improvements to the systems, has now been reduced to the much lower level of 0.0005 or a chance of an error of 1 in 2000 tasks. The total error probability is now 0.0025 or a chance of 1 in 400 tasks. This of course is now better than the random error probability of 1 in 500 which by definition is not affected by the system improvements. The contribution of the random element is now 80 per cent of the total error probability. Obviously further system improvements would not be greatly beneficial in reducing the total error probability, since this could never be reduced below the residual error probability of 1 in 500 due to the random element. This is probably a rather extreme case but it has been used to demonstrate the principle of residual random error showing that it is impossible to eliminate human error completely.

1.2.6 Human error and risk

The concept of residual error is important when the contribution of human error to the risk of certain activities is considered. It is frequently stated that the risk of an activity can never be reduced to zero, but can hopefully be reduced to a level which is considered acceptable when weighed against the benefits of the activity. It is also a fact that the cause of about 80 per cent of all accidents can be attributed to human error. The fact that human error cannot be entirely eliminated must therefore have an important bearing on the level of residual risk of an activity where human error is a potential accident contributor. Nevertheless, in such activities, the opportunity to reduce risk to acceptable levels by reducing the probability of systemic errors always remains a possibility. The main theme of this book is to identify some of the more common deficiencies which are found in systems and which make human errors more likely.

References

Hollnagel, E. (1993). *Human Reliability Analysis: Context and Control*, London: Academic Press, p. 29.

Meister, D. (1966). Human factors in reliability. In W.G. Ireson (ed.) *Reliability Handbook*, New York: McGraw-Hill.

Reason, J. (1990). *Human Error*, Cambridge: Cambridge University Press, p. 9.

Swain, A.D. and Guttman, H.E. (1983). *Handbook of Human Reliability Analysis with Reference to Nuclear Power Plant Applications*, Washington DC: U.S. Nuclear Regulatory Commission, pp. 2–7.

2

Errors in practice

2.1 Introduction

Human error can occur in a wide variety of types and forms and be manifested in an almost infinite number of ways. To introduce some order into what might otherwise become a veritable forest of errors, attempts have been made by experts in the field to formulate ways of organizing errors into taxonomies. The definition of a taxonomy, according to *Collins English Dictionary*, is 'a grouping based on similarities of structure or origin'. Taxonomies are usually applied in the biological sciences to classify animal or plant genera and species and to provide a framework of orders, families and sub-families. The purpose of a human error taxonomy is exactly the same. However, as with animal or plant species, there are many different ways of organizing human errors into taxonomies. The choice of taxonomy depends mainly on the purpose for which it is to be used. A number of different taxonomies are described below.

2.2 Genotypes and phenotypes

2.2.1 Definition

Perhaps the broadest taxonomy is the division of human errors into phenotypes and genotypes (Hollnagel, 1993). As with the use of the term 'taxonomy' to classify human errors, the words 'phenotype' and 'genotype' are also taken from the biological sciences. A genotype is defined as the 'internally coded, inheritable information' carried by all living organisms (Blamire, 2000). This stored information is used as a 'blueprint' or set of instructions for building and maintaining a living creature. A phenotype is defined as the 'outward, physical manifestation' of an organism. These are the physical parts of the organism, anything that is part of the observable structure, function or behaviour of a living organism (*op. cit.*).

In the field of human error, the genotype of an error similarly relates to its origin. For example a task to be carried out may have to be selected from a number of possible tasks. If an error occurred during the mental selection process, such that the selected task was inappropriate, then even though the task was carried out correctly it would lead to an undesirable outcome. Such situations arise when unfamiliar or complex tasks need to be carried out, requiring conscious thought and careful planning. The source or genotype of the error therefore lies with the selection process and not with the implementation of the task.

By contrast, the phenotype of an error, as with the biological definition, is concerned with its physical outward manifestation when the genotype is eventually translated into observable action. The phenotype is how an error is physically revealed in a faulty action (or lack of action) that, like the physical manifestation of an organism, is fully observable and, in the case of an error, possibly measurable. This is completely unlike the genotype of the error which by contrast cannot be outwardly observed or easily measured, except perhaps if it is later remembered by the person carrying out the task when the undesirable outcome is observed.

In simple terms, the difference between the genotype and phenotype of an error reflects the distinction between cause and effect. However, while the genotype relates to a cause at the cognitive level, the phenotype relates to the effect at the observable level of execution of the task leading to an undesirable or unwanted outcome. As stated in Chapter 1, it is only when the actual outcome is compared with the desired outcome that it becomes known an error has been made. But the phenotype of the error is always the faulty action, not the outcome, the outcome only triggers the realization of an error. As with a living organism, if a mistake occurs in the genetic coding then the organism that results will be a mutated version of what would normally be expected and possibly non-viable. However, the coding error will only become obvious when the organism is observed (in the absence of genetic testing). Faulty 'coding' in the cognitive process similarly results in a 'mutated' task identified by an undesirable outcome. When an observed error (the phenotype) results from faulty cognitive processing (the genotype) then a psychologist might be more interested in the latter (i.e. in what went wrong within the cognitive process). An engineer would be more interested in the phenotype and in whether there was an external cause that might have led to faulty cognitive processing. As already stated, causes can usually be traced to deficiencies in the systems that can influence task performance. From an engineering perspective, information about the systemic causes of an error is much more useful than knowledge of the faults that occurred in mental processing, since the former may be corrected. However, knowledge of the latter could also be useful in some cases.

2.2.2 Example of phenotype/genotype taxonomy

The phenotype/genotype taxonomy can best be explained using a practical example. The taxonomy is set out diagrammatically to the left of Figure 2.1 and is illustrated on the right by the example of a visual inspection task. The task involves an inspector, who is trained visually to recognize surface defects in small metallic components

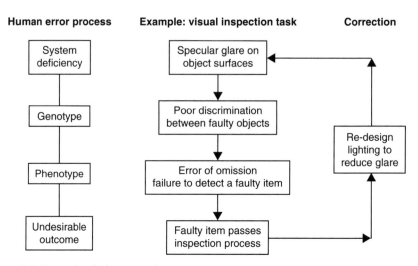

Figure 2.1 Example of phenotype/genotype taxonomy.

which pass along a conveyer belt. Due to the position of the overhead lighting, the metallic items can be subject to specular glare (i.e. they reflect light back to the observer) obscuring surface defects. As a result the inspector may allow faulty items to slip past and enter the product chain rather than being removed and discarded. The genotype of this error is that the inspector is unable to discriminate sufficiently between acceptable and sub-standard items. This is due to a normal limitation in acuity and therefore mental processing. The phenotype is the actual observed error, a failure to detect an item with surface faults. In this case, both the phenotype and genotype are of interest to the engineer who is charged with effective operation of the production line. The 'undesirable outcome' indicates that there is a problem to be investigated. The 'phenotype' is identified as the observable and immediate reason for the problem. However, knowledge of the phenotype alone, 'a failure to detect an item with surface faults', is not sufficient to correct the problem. The genotype of the error needs to be investigated. Through observation of the production process and by discussing the problem with the inspector, it is possible to arrive at an explanation of why the errors are being made, logically tracing the problem back to its ultimate source, the defective lighting system. The engineer can then correct the system fault.

While this is a rather simplistic example, the method of analysis organizes the known information in a logical way so that the steps needed to resolve the problem can be identified. Figure 2.1 takes the analysis a step further and includes on the far right of the diagram a correction loop to overcome the problem. The analysis of the example is set out diagrammatically in a top down hierarchical sequence, commencing with the system deficiency and moving downwards to the undesirable event. Closing the loop is then a bottom up process commencing with the most observable and immediate effect, the undesirable event. It is a process of detection of the problem (the phenotype), diagnosis of the cause (the genotype) and correction (re-design the lighting system). This approach to problem analysis ensures that each step is considered logically without any omissions, and enables the process to be documented.

The genotype of the error is closely associated with the interaction between the user and the system at the sensory or input level. The phenotype, however, is associated with the interaction between the user and the interface at the action or output level. The term user system interface (USI) is often used to describe this interaction and in this context is synonymous with the human machine interface (HMI) which is often referred to in human factors literature.

The usefulness of an error taxonomy depends on the context in which it is to be used. Methods of representing human error are described in later chapters and it will be seen that different contexts will demand different taxonomies, or indeed combinations of these. Thus it cannot be said that any particular taxonomy is more useful than another without knowing the context in which it is to be used. The phenotype/genotype taxonomy, although not widely used, can be useful in the investigation of systemic errors at a fairly simple level.

2.3 The skill, rule and knowledge taxonomy

2.3.1 Definitions

The skill, rule and knowledge (SRK) based taxonomy was developed by Rasmussen (1983) and has since been widely adopted as a model for describing human performance in a range of situations. The three categories of SRK behaviour as proposed by Rasmussen are defined below and each of these are illustrated by a simple example. The Rasmussen taxonomy is also useful because once a task has been classified, it is possible to estimate the reliability of the performance of the task within a range of human error probability. As performance moves from skill based through rule based to knowledge based, the reliability tends to decrease. Typical error probabilities for each behaviour type are also suggested below.

2.3.1.1 *Skill based behaviour*
2.3.1.1.1 Definition
Skill based behaviour represents the most basic level of human performance and is typically used to complete familiar and routine tasks that can be carried out smoothly in an automated fashion without a great deal of conscious thought. Indeed the tasks that can be carried out using this type of behaviour are so familiar that little or no feedback of information from the external or work environment is needed in order to complete the task successfully. It is effectively carried out under 'automatic pilot' and is said to exhibit a high degree of feed-forward control. This type of behaviour is generally highly reliable and when errors do occur they are either of a random nature or occur because of interruptions or intrusions into familiar patterns of activity.

2.3.1.1.2 Error probability
A typical range of error probability for skill based tasks is from as high as 0.005 (alternatively expressed as 5.0E−03 or 5×10^{-3}) or 1 error in 200 tasks to as low as 0.00005

(5.0E−05 or 5×10^{-5}) or 1 error in 20,000 tasks on average (Hannaman and Spurgin, 1983). The expression of probability values in exponential units (e.g. 5.0E−03 or 5×10^{-3}) is described in Chapter 4.

2.3.1.1.3 Example – car driving

The mechanical process of driving a car is an everyday example of skill based behaviour. An experienced driver will commit the highly routine tasks of steering, changing gears, clutch operation and braking to what is effectively 'automatic pilot' in order to concentrate on the slightly more complex task of navigating to the desired destination. One has only to recall the early stages of learning to drive to appreciate how little conscious attention needs to be given to these tasks once driving experience has been gained.

2.3.1.2 *Rule based behaviour*

2.3.1.2.1 Definition

Rule based behaviour is adopted when it is required to carry out more complex or less familiar tasks than those using skill based behaviour. The task is carried out according to a set of stored rules. Although these rules may exist in the form of a set of written procedures, they are just as likely to be rules that have been learned from experience or through formal training and which are retrieved from memory at the time the task is carried out. As with skill based behaviour, rule based behaviour is also predominantly feed forward in nature, the rules being applied with minimal feedback from the situation except perhaps for the detection of 'waypoints' to indicate the progression of the rule based procedure or sequence of actions.

2.3.1.2.2 Error probability

Error probability values for rule based tasks are typically an order of magnitude higher than for skill based tasks. They lie within the range from 0.05 (5.0E−02 or 5×10^{-2}) or 1 error in 20 tasks to 0.0005 (5.0E−04 or 5×10^{-4}) or 1 error in 2000 tasks on average (Hannaman and Spurgin, 1983).

2.3.1.2.3 Example – car driving

Successfully driving a car along a familiar route, say on a daily journey to and from work, is an everyday example of rule based behaviour. All car drivers will have experienced the rather alarming sensation of arriving at a familiar destination by car and not being able to recall certain segments of the journey. This is in spite of carrying out driving manoeuvres such as stopping at traffic lights and negotiating roundabouts while taking account of other traffic. This is only possible because navigation on familiar journeys is undertaken by automatic retrieval of a stored rule set based on a mental map of the journey.

2.3.1.3 *Knowledge based behaviour*

2.3.1.3.1 Definition

Knowledge based behaviour is adopted when a completely novel situation is presented for which no stored rules, written or otherwise, exist and yet which requires a plan of action to be formulated. While there is clearly a goal to be achieved the method of achieving it will effectively be derived from first principles. Once a plan or strategy has been developed, this will be put into practice using a combination of skill and rule based actions, the outcome of which will be tested against the desired goal until success is achieved. Since this may involve a process of trial and error in order to optimize the solution, knowledge based behaviour involves a significant amount of feedback from the situation. However, with experience and learning, what was initially classed as knowledge based behaviour, may ultimately become more rule based. Conversely, it may be found that a task previously carried out using rule based behaviour will need to be undertaken using knowledge based behaviour if it can no longer be successfully completed using the current set of rules. This may happen when the rules are no longer applicable because of some change in the situation.

2.3.1.3.2 Error probability

Knowledge based tasks have significantly higher error probabilities than either skill or rule based tasks mainly because of the lack of prior experience and the need to derive solutions from first principles. Error probability values vary from 0.5 or 1 error in 2 tasks to 0.005 (5.0E$-$03 or 5×10^{-3}) or 1 error in 200 tasks on average (Hannaman and Spurgin, 1983).

From an examination of the range of probability values for each of the three SRK types of behaviour, it can be seen that some overlap exists across the stated ranges. While the probability of error in knowledge based, behaviour can be as high as 1 in 10 tasks carried out, at its best the probability can equal the lowest probability for skill based behaviour which is normally considered to be quite reliable. It should also be noted that the selection of a probability value for a particular task from the given range for the behaviour type will depend upon the capability of the person undertaking the task and the favourability of the external performance shaping factors.

2.3.1.3.3 Example – car driving

While driving a car on a familiar journey much of the journey will be conducted using a combination of skill and rule based behaviour as described above. However, if the demands of the driving situation increase, for example in negotiating unfamiliar roads or new and complex traffic layouts, or when meeting a novel situation such as a diversion because of an accident or hold-up, the driver may need to switch to knowledge based behaviour. In this situation, the journey is no longer on 'automatic pilot' and the driver may need to work out new waypoints for his route from first principles. In practice, because much of the process of car driving has a high degree of familiarity, a car driver will continuously switch between S, R and K based behaviour according to the situations encountered and the skill and experience of the driver in relation to these. A similar switching strategy will be used in many industrial tasks.

2.3.2 Selection of appropriate behaviour type

2.3.2.1 *Principle of cognitive economy*

While the three types of behaviour are distinctive, it can be seen from the example of car driving that in many task sequences the behaviour will oscillate continuously between the three types depending on the characteristics of the task and the capability of the person carrying it out. The underlying purpose of this is best understood by reference to the principle of 'cognitive economy' (Reason, 1990). This principle is based on limitations in mental processing (i.e. cognitive) resources which restrict their capacity to:

(a) assimilate external information,
(b) understand its implications for the current situation in terms of appropriate actions to produce a desired outcome,
(c) carry out these actions.

This full process when carried out in its entirety is equivalent to Rasmussen's category of knowledge based behaviour and is extremely time and resource intensive. While such a resource intensive operation is being carried out, the capacity effectively to undertake parallel tasks, even at a simple level, is severely compromised, and is likely to lead to an increased probability of error. Thus, it is always more economic and effective, when faced with any slightly complex situation requiring action, to fall back on a set of stored rules that have been acquired from past exposure to similar situations. A fairly rapid process of situation matching can then be carried out to select the most appropriate set of rules for the situation and apply these. What would have otherwise been classed as knowledge based behaviour is therefore relegated to the rule based category.

2.3.2.2 *Situation matching*
The effectiveness of cognitive economy in reducing demands on limited resources is clear. However, the potential danger of adopting the rule based approach is that its success is highly dependent on the accuracy of the situation matching process so that the most applicable rules are selected. Where situations are highly familiar and have been encountered on many previous occasions, then the matching process is likely to be accurate, the rule selection appropriate and the cognitive economy justified. Where the similarity between the existing and past situation is less clear-cut, then it is more likely that a faulty rule selection will be made. The natural desire for economy of effort in many workplace situations, perhaps due to work overload, increases the likelihood that rule based behaviour will be adopted where knowledge based may be more appropriate. The probability of a rule based error may then increase over and above that of a knowledge based error. The type of error that can result from a faulty rule selection is in a category of its own and is often referred to as a 'mistake' (Reason, 1990). This is further discussed in Section 2.4.3.

2.3.2.3 *Continuity of attention*
In a similar way to the relegation of knowledge based to rule based behaviour, the adoption of skill based behaviour enables an even greater reduction in attentional resources to be achieved. As discussed above, skill based tasks take place in a highly automated mode and thus they clear the way for tasks requiring more conscious attention and/or cognitive processing power. However, this form of economy has its own peculiar drawbacks. While skill based tasks can be highly reliable, since they are based on highly practised routines, they are extremely vulnerable to interruption and interference

(Reason, 1990). Their reliability seems to depend heavily upon maintaining a high degree of continuity of the automated process without the significant intervention of conscious thought or attention. Indeed such intervention can wreak havoc with the learned routine and completely degrade performance. It is, however, possible to limit the effects of inter-ference or interruption of skill based behaviour when different senses are involved. For instance, it is easily possible to drive a car, a task where the input is mainly visual, with minimal distraction while at the same time speaking to a passenger. On the other hand, attempting to select a radio station or operate a mobile phone while driving involves a serious visual distraction which is likely to increase reaction time in the event of a sudden demand due to a change in the road conditions. This is in addition to the hazard caused by removing one hand from the steering wheel.

2.3.2.4 *Human error in the learning process*

Interference or interruption of the cognitive processes in skill based behaviour can increase the probability of human error in many situations, but it is particularly dis-ruptive of human performance in situations involving novices. This is most easily illustrated by the common example of learning to ride a bicycle. In the early learning stage of this process, the novice rider will need to apply a great deal of conscious effort to maintain balance and direction, both essential to successful performance. The process approximates to what in cognitive terms would be a knowledge based activity but using motor, rather than cognitive, skills. As the activity becomes more practised, the process becomes semi-automated and more skill based requiring less and less con-scious thought and attention. At this point the learner will become more adapted to the changing situation ahead and direction and stability will become better integrated. With care the novice will be relatively safe to share the road with other users. Yet in these early stages, the process of riding a bicycle is still a learning process rather than an accomplished skill. The danger is that this may not be fully appreciated by the novice and the problem of over-confidence arises in what might seem to be an effort-less routine. On suddenly being presented with an unfamiliar situation, it becomes clear that a degree of conscious effort is still needed to maintain balance and direction. The sudden call upon attentional resources by the new situation suddenly degrades riding performance and an accident becomes more probable.

A similar principle applies in the case of young novice car drivers, whose accident proneness following their first acquisition of a licence is reflected in the level of insur-ance premiums demanded from them. The same danger exists in many workplace situ-ations where the early stages of learning appear to have been fully absorbed and a degree of proficiency obtained. As with the bicycle rider, the danger then becomes one of over-confidence.

There is also a philosophical aspect to achieving very high reliability of automated routines carried out without conscious effort or intervention. Readers interested in this aspect of skill based performance are referred to the book 'Zen in the Art of Archery'. This book describes how Zen masters help their students to learn through an 'ego-less' experience rather than through lecturing or repeated demonstrations. This can, accord-ing to the author, lead to a standard of proficiency which can only be destroyed through a conscious or egocentric attention to technique (Herrigal, 1999).

Table 2.1 Determination of behaviour type

	Criteria			
Type of task	Routine task?	Fully understood?	Well practised?	Procedure required?
Skill based	Yes	Yes	Yes	No
Rule based	Yes	No	No	Yes
	No	Yes	No	Yes
Knowledge based	No	No	No	No procedure available

2.3.3 Determination of behaviour type

In order to classify a task within one of the three categories in the SRK classification it is useful to summarize the general criteria that need to be met for a behaviour type to be applicable. A suggested set of criteria is given in Table 2.1.

It should be noted when using this table that the requirement for a 'procedure' can include either written procedures or mentally stored rules, the requirement being determined by the complexity of the task. From the table it is seen that the requirement for skill based tasks is that they are always routine, fully understood and well practised. Rule based tasks can also be routine in nature, but because they are less frequent or familiar than a skill based task they will generally not be well practised and/or not fully understood and a procedure will therefore be required. If a task is not routine, but is fully understood, then again a procedure will be required and the task can be defined as rule based. If a task is not routine, not fully understood and a procedure is not available then this must be defined as a knowledge based task. The table is not meant to be completely definitive but gives an indication as to the category of behaviour into which a particular task may be fitted.

The SRK taxonomy has been widely used as the basis of a number of other taxonomies and techniques for understanding human error. The generic error modelling system (GEMS) taxonomy (Reason, 1990), described below, was based on the SRK approach and represented an important step forward in classifying human error.

2.4 The generic error modelling system taxonomy

2.4.1 Overview

The GEMS (Reason, 1990) adopts a hierarchy based on the Rasmussen SRK taxonomy. Errors within the skill based category of tasks are further sub-divided into 'slips' and 'lapses' both of which are errors which take place during execution of a task and comprise a failure to achieve the desired outcome from an action. Errors that take place within rule and knowledge based tasks are classified as 'mistakes' and these correspond to errors made in the planning rather than the execution process. Definitions of each type of error in the GEMS taxonomy are as follows:

● A *slip* is a failure in the execution of an action as planned.

- A *lapse* is an omission to execute an action as planned due to a failure of memory or storage.
- A *mistake* is an error in planning an action irrespective of whether the action is carried out correctly. This is further sub-divided into two types of mistake:
  - a *rule based mistake* occurs during the selection of a plan to achieve a desired goal. It can take the form of the misapplication of a good plan (from a set of stored plans based on similar or previously experienced situations) or the application of a bad or inadequate plan (that is a plan that was not adequate to achieve the desired goal). Reason refers to this type of mistake as a 'failure of expertise'.
  - a *knowledge based mistake* occurs during the formulation of a novel plan to meet a situation for which no plans currently exist. It is essentially a failure in reasoning due to inadequate or inaccurate knowledge of the situation or an inability to combine different facets of accurate knowledge in a way that will lead to appropriate action. Reason refers to this type of mistake as a 'lack of expertise'.

Reason (1990) addresses the many and various reasons why slips, lapses and mistakes may occur showing how, in a sequence of tasks, a person will move between S, R and K based performance according to their own perceived level of skill and confidence and the unfamiliarity or complexity of the task. Each of the above types of error is now discussed in more detail.

2.4.2 Slips and lapses

Slips and lapses occur at the skill based level of performance which, as described above, is applicable to highly routine and familiar tasks. When slips and lapses occur this will be frequently due to:

(a) Lack of attention at points during the task that needs to be monitored as a check upon progress to completion. The lack of attention may be brought about by:
  - the conscious mind drifting to unrelated matters perhaps because the task is so routine that there is a loss of vigilance or arousal leading to boredom;
  - external events intervening such as the requirement to answer a telephone, or enter into a conversation.
(b) Over-scrupulous attention to points in the task that do not need to be monitored. This introduces a distraction into a sequence of actions that would normally be carried out automatically with little or no conscious thought.
(c) Lapses in particular may occur when, after an interruption the procedure may be recommenced further along the sequence, having inadvertently omitted one or more tasks. Alternatively, it may be that an intention to begin a task is delayed, perhaps, as a result of waiting for some system state to be achieved. The delay results in other demands being made upon conscious thought which cause the start signal for the task to be missed.

Clearly, there are many possible reasons for slips and lapses and it is only possible within the scope of this book to mention a few of the most common reasons. However, a knowledge of why slips, lapses and mistakes occur is useful in the human error analysis of accident case studies presented in Part II of this book since it may help to identify system deficiencies which may have precipitated the errors.

2.4.3 Mistakes

The success of rule based behaviour is, as described above, highly dependent upon accurate pattern matching. If a set of rules has been successfully applied in the past to situations that are similar to that currently presented, then a comparison is made between the two situations. Points of similarity are identified and if these are considered numerous enough or the similarity strong enough, then an attempt is made to apply the previous set of rules to the new situation. The success of the rule application is monitored by assessing the degree to which the intended outcome is achieved. Of course, there may well be more than one set of rules that are potentially applicable to the new situation. In this case, alternative sets of rules will be ordered in a mental hierarchy, the set having the highest perceived success probability being placed at the top of the list.

Mistakes at the rule based level can, as described above, be attributed to either the *misapplication of good rules* or the *application of bad rules* (Reason, 1990) and this is further developed below.

2.4.3.1 *Rule based mistakes*

2.4.3.1.1 Misapplication of good rules

The rules are considered to be *good* where they have been proven in circumstances that are similar to the current situation but differ perhaps in a minor but important respect that has not been recognized. Reason uses the description *strong-but-wrong* to describe a compelling factor that would influence the choice of a rule that had successfully been used before. A rule may be strong but wrong due to a high frequency of previously successful use of the rule or because the rule has been used quite recently (i.e. there is a strong memory of a successful application). The strong experience is carried forward into the present and influences current behaviour. The main point to be made is that there is nothing wrong with the rule that is selected since it has worked perfectly in the past. The problem lies with the fact that the rule is not applicable to the present situation.

A major difficulty which arises with the strong-but-wrong type of behaviour is that of perseverance of the belief that the rule is applicable in spite of compelling evidence to the contrary. This tendency to stick with the familiar, rather than branch out into unknown territory, and apply rules which may be less familiar but more applicable, has in the past led to a number of major accidents, some of which are described in Part II of this book. This can result from system deficiencies that result in misleading information being given to the user resulting in the wrong application of a familiar rule.

2.4.3.1.2 Application of bad rules

Bad rules are rules which, at best, are less than adequate when it comes to providing an acceptable solution to a problem, although they may provide some sort of solution. At worst, the rules are intrinsically defective and will not provide a solution under any circumstances. The rules may be bad because they do not fully encapsulate the requirements needed to solve the problem or it may be that some aspect of the problem has been over-looked, or has not been fully appreciated, making the selected rule non-applicable. The difference between good rules (as defined above) and bad rules is that with bad rules there is no relevant previous context, that is, no support from a previous application of the rule in a similar situation that would make it effective in this situation.

Of course, the root of the problem is not the bad rules, it is the flawed logic used in assessing the applicability of the rule to the situation, possibly due to a lack of expertise or an over-optimistic view of one's own expertise. If the situation is not fully understood, then the reasoning used to select and apply the rule will be based on a false premise, mak-ing the rule a bad rule as far as this situation is concerned. When seen in this light, rule based mistakes are difficult to distinguish from knowledge based mistakes (discussed below) due to the degree of novel thinking involved in rule selection. The distinction becomes clearer when it is realized that the application of a bad rule unequivocally requires some sort of rule (defined as a *predefined* plan of action) to be brought into con-sciousness and applied no matter how wrong or inapplicable that rule might be. Knowledge based mistakes do not involve the selection of prior sets of rules at all.

2.4.3.2 *Knowledge based mistakes*

In some ways, mistakes at the knowledge based level are easier to comprehend than the rule based mistakes considered above. It is difficult to visualize the cognitive and logi-cal processes involved in selecting an appropriate rule from a set of predefined rules, especially when this appears to take place smoothly and rapidly, without a great deal of conscious effort. By contrast, the slower and more laborious application of knowledge based behaviour to solve a novel problem, is the main focus of consciousness, to the exclusion of any other serious cognitive activity until a solution is found. This process is quite easy to visualize. It takes place in a number of stages and comprises:

1. assimilation of information about the problem based on the observable facts,
2. formulation of a mental model of the problematical situation,
3. manipulation of the model to construct hypothetical scenarios which are used to visualize possible outcomes,
4. selection of a preferred problem solving strategy and translation of this into action.

During this process, knowledge based mistakes can occur which produce a faulty strategy that, although implemented as planned, will not produce the desired solution. These mistakes tend to originate from two possible sources.

2.4.3.2.1 The person

Inadequacies arise from poor internalization of the problem as a model that can be mentally manipulated in the form of alternative scenarios due to:

- *Inability to see the greater picture*. Limited cognitive processing capacity makes it impossible to focus on and bring together all the facets of the situation requiring attention to find an optimal solution.
- *Selective focusing* on facets of the situation that are more familiar or better understood.

- *Excessive emphasis* on the importance of facets that more easily come to mind or are more recent in memory.
- *Disregarding contradictory evidence.* Once a strategy has been formulated and selected, evidence that contradicts this tends to be ignored in favour of evidence that supports it.
- *Tendency to haste* in selecting and implementing a strategy in order to impose order on chaos and reduce anxiety, perhaps not giving adequate consideration to alternative strategies. The same haste is also applied to checking processes used to confirm that the correct strategy has been selected. This also makes it more likely that contradictory evidence will be ignored.

Reason suggests a number of other behavioural tendencies that hinder the selection of a correct problem solving strategy. These include *thematic vagabonding* (flitting from one strategy to another without investing sufficient time or effort in any one strategy to make a reasoned selection) and *encysting* (being the opposite of *thematic vagabonding*, the tendency to dwell at length on one particular strategy to the exclusion of alternative approaches).

The result of defective internalization of the problem is that the mental model of the situation that is developed is a poor 'fit' against reality. This will be revealed following or during implementation of the action based on the selected strategy. Even though the action is carried through correctly, the results will not meet the requirements of a successful solution.

2.4.3.2.2 The situation

Problems arising from the situation usually result from inadequate or misleading information available from the external environment thus preventing the formulation of a successful strategy (given that the cognitive processing power of the problem solver is adequate). Lack of information may be due to:

- *Inherent opaqueness* of the situation resulting from an absence of stimuli or cues which could otherwise lead to successful strategy formulation.
- *Defective presentation* of information that is available and would otherwise be adequate to formulate a correct strategy. Since this situation is potentially correctable, any mistake made would then be classed as a systemic error.

Inadequate or misleading information resulting from the above will be present in many novel or unfamiliar situations. The only option remaining may then be a trial and error approach, undesirable as this may be. Trial and error may, at its worst, equate to the direct application of random strategies to increase the amount of information available, the worth of a strategy being assessed in terms of the relative success of the different outcomes. The trial and error approach would clearly be unacceptable in some situations where a wrong outcome has negative safety implications. In hazardous processes in particular, the design should if possible allow the system to be brought to a safe condition through a single action, in effect a 'panic button' such as the 'scram' system on a nuclear reactor.

Knowledge based mistakes have been shown to be implicated in a number of serious major accidents. In some cases, a little foresight might have provided more and/or better presented information to guide the user towards a successful resolution of the problem. However, there will always remain novel situations whose occurrence is virtually impossible to foresee and some of these may result in serious accidents. Such accidents are notoriously difficult to predict simply because the need to invoke knowledge based behaviour implies that no applicable set of rules has been developed to cope with the situation. There will always be situations of this kind. However, to end on a less pessimistic note, it is a tribute to human ingenuity that on so many of these occasions, the human operator has risen above all the adversities listed above, and developed a novel solution which fully meets the needs of the moment.

2.4.4 Summary

The GEMS taxonomy described above is extremely popular among human error practitioners because within its scope it is able to capture most manifestations of human error. It is not only valuable as a descriptive tool, but it also allows for the possibility of error correction. Once an observed human error has been placed within the taxonomy, it is possible by reference to GEMS to identify ways in which the cognitive mechanisms may have failed. By exploring the context of the error (the environment in which it occurred) it may be possible to pinpoint system deficiencies which may have been error contributors. In turn, some of these may be capable of correction. The second part of this book deals with case studies of accidents caused by human error. The main purpose of this is to explore whether a better knowledge of the interaction between the person making the error and the system that may have induced the error, could have allowed the accident to be avoided. In turn, it is hoped that the lessons learned from the study of past accidents may reduce the frequency of similar (or even dissimilar) accidents in the future.

References

Blamire, J. (2000). http://www.brooklyn.cuny.edu/bc/ahp/bioinfo/GP/definition.html.

Hannaman, G.W. and Spurgin, A.J. (1983). *Systematic Human Action Reliability Procedure (SHARP)*, Palo Alto: Electric Power Research Institute, p. A-8.

Herrigal, E. (1999). *Zen in the Art of Archery*, London: Random House.

Hollnagel, E. (1993). *Human Reliability Analysis: Context and Control*, London: Academic Press.

Rasmussen, J. (1983). Skills, rules, knowledge: signals, signs and symbols and other distinctions in human performance models. *IEEE transactions on systems man and cybernetics*, SMC-13, 257–267.

Reason, J. (1990). *Human Error*, Cambridge: Cambridge University Press.

3

Latent errors and violations

3.1 Introduction

Latent errors and violations are considered together for convenience in this chapter, although strictly speaking they are unrelated. When an error is said to be latent it simply means that a significant time has elapsed between an error being made and the existence or consequence of that error being manifested. When an error is not latent it is said to be active. An active error is an error where the effects are manifested immediately or almost immediately. Active errors are further discussed below and case studies for active errors are presented in Chapters 10 and 11 of this book.

The concept of latent error is crucial to an understanding of human error and accident case studies involving latent error are described in Chapters 7–9 of this book. Latent errors are potentially accidents waiting to happen. The key to preventing such accidents lies therefore in the ability to detect latent errors as soon as possible after they have occurred. The longer a latent error is allowed to exist the less likely that it will be discovered and the more likely that it will cause an accident. The most common areas of activity where latent errors take place are within maintenance and management, and the distinctive features of each of these types of latent error are discussed below. Rather than relying upon chance to discover latent errors, or even worse an accident, which is often what happens, it is better to formulate a strategy for detection. In the case of maintenance errors, the strategy involves checking for latent errors after tasks have been carried out, and in the case of management activities to instigate some sort of independent review or audit.

Violations comprise a special class of human error, and indeed some human error specialists would not even classify these as errors at all. A violation is effectively an action that has taken place in breach of a set of normal operating rules, whether or not these rules are written down, are implicit within the action or have been developed as part of custom and practice. The main difference between a violation and an error is that a violation is often carried out with conscious intent, whereas an error by definition takes place unintentionally. Conscious intent is almost always present with violations which are one-off events defined as 'occasional violations'. However some violations, as discussed in Section 3.3, are 'routine violations', and indeed the person making the violation may come to regard this way of working as normal practice. In this sense the degree of conscious intent may diminish with time and it may eventually be consistently forgotten that a rule is being violated. This could occur for instance, in the case of an action carried out in breach of rules that have either fallen into disrepute or are no longer enforced by management or obeyed by the workforce. The strategy for detecting and correcting violations is usually some sort of quality audit or management review of the way activities are carried out. A number of accident case studies involving violations are described in Chapter 12.

3.2 Latent and active errors

3.2.1 Introduction

One of the most common environments where latent errors occur is in maintenance activities. Latent errors made during maintenance activities may not be revealed until the system is put back into service with possibly catastrophic results. Such an error is revealed through a fault in operation, which may or may not occur immediately after the system is re-started following maintenance. The other important area where latent errors occur is in management decision-making. In this respect, it is necessary to make a distinction between high-level and low-level decision-making. Minute-by-minute decisions that are made at the lower levels of an organization, for instance, at the operational level of a processing facility, are often followed by immediate action so that any errors that are made become quickly obvious from their consequences. These might then be defined as active rather than latent errors. However, the effects of high-level management decisions may not always be immediately apparent since they need to be propagated downwards through the organization before the consequences are seen. This is usually because the point in the organization at which the decision is made is remote from, and only indirectly connected with the situation where its effects will be manifested.

3.2.2 Active errors

Active errors become immediately obvious at, or soon after, the time the error is made. The reason for this is the close association between the task and the error. Usually an active error is revealed through some sort of feedback to the person carrying out the task. The feedback is that:

(a) the outcome of the task is not as desired, indicating that an error has been made or
(b) the error itself is detected before the undesirable outcome is noticed.

Active errors are more common in operating situations such as the use of manual equipment, hands-on control of machinery, driving a vehicle, etc. What defines this type of task is the immediacy of the feedback from the task to the operator such that if an error is made it is quickly obvious. In addition, tasks involving active errors tend to be complete within themselves rather than forming a part of a sequence or procedure where success is measured on completion of the procedure, perhaps some considerable time later.

It is quite common that when a task forms part of a longer sequence, an active error is revealed because the sequence cannot be carried forward beyond the point where the error has occurred. This triggers recognition of the error. A simple domestic example of this situation would be entering a darkened room to retrieve an object. In switching on the electric light for the room, it is necessary to select the correct switch (from a set of external switches illuminating different rooms) and then operate the switch that has been selected. This task is a small but important part of the broad activity of retrieving the object from the room. In this case, detection of the error does not depend on the completion of the wider activity. Until the correct light switch is thrown, the wider activity cannot be progressed any further (except with difficulty). If the wrong switch is selected and operated, then the light in the room to be entered does not come on and the selection error is immediately obvious. This is typical of an active error. It is also an example of an error which has an inherent recovery mechanism built into it. Error recovery is a topic that is dealt with separately in Chapter 5.

3.2.3 Latent errors

3.2.3.1 *Latency*

The immediacy of active errors usually enables their rapid discovery either through detecting that an error has been made or detecting the consequences (in terms of an undesired outcome) immediately after the error has been made. By definition, the consequences of a latent error are not immediate due to the inherent latency. The immediate discovery of a latent error is therefore only possible when the error is made since the consequences are by definition delayed. If the consequences can lead to an accident then the only method of preventing that accident is to discover the latent error at the time it occurred. In the case of active errors, it may not be possible to complete a sequence of tasks without the error being revealed, as described in the example of the light switch above. However, if a latent error occurs it will by definition be possible to complete the sequence of tasks without the error being discovered.

A domestic example, the making of a cup of tea, can be used to illustrate latent error. An action that takes place in the early stages of tea making is the pouring of milk into the cup. It is possible that the error of failing to pour milk into the cup would be detected immediately it occurred, in which case it could be classed as an active error. However, it is more likely that the error would not be noticed until after the tea had brewed and was poured into the teacup. The colour of the tea would reveal the error, the main point being that the procedure is complete before the error is discovered. By this time the error is no longer recoverable, unless of course it is decided to pour milk into the tea. For dedicated tea makers this would be regarded as a failure of the procedure! That is a rather trivial example but later in this book there will be found many examples of more serious situations where there is no opportunity for a latent error to be discovered or recovered once it has been made.

The time period or *latency* before latent errors are discovered can vary widely and in the context of design activities or management decision-making, the time periods can extend to months or even years. In the case of maintenance activities the latency will usually be much less than this and it may be possible to engineer a method of feedback which reveals the error, rather than wait for the consequences of the error to occur. This feedback often takes the form of checking or testing whether a task has been carried out correctly, for instance by test running a piece of machinery in order to demonstrate correct operation. The objective is to pre-empt a fault occurring later, when the machinery is required to operate flawlessly and the opportunity for error correction will not exist. However, in some cases the time window or opportunity within which this can be done is quite small. In the tea-making example, the outcome might be pre-empted if it could be checked that there was milk in the cup, but this must take place before the tea is poured. If the error was then discovered, the milk could be added to the cup before the tea was poured rather than *vice versa*.

3.2.3.2 *Discovery of latent errors*

The hidden nature of latent errors would appear to make them more insidious than active errors. However, they also have the advantage of being potentially more recoverable since there will be an opportunity for discovery and correction before their effects are manifested. The difficulty is that many latent errors, once they have occurred, are extremely difficult to detect. Discovering latent errors which could have serious consequences for a maintenance procedure, for example, would involve a review of all the potential errors which might have occurred during the procedure. This can be extremely time consuming and it is never certain that the prediction of errors would be sufficiently thorough to identify all errors with dangerous consequences. However, this is not always necessary and maintenance errors are best discovered through proof testing of maintained equipment, if the equipment develops a fault then this will reveal the error. There are two drawbacks to this:

- Equipment may have to operate for a considerable time before the error is revealed and this may not be possible or convenient.
- The error could reveal itself by the undesirable outcome that the test is meant to pre-empt. For instance, the error of failing to put lubricating oil in a gearbox would result in damage to the gearbox through lack of lubrication when the equipment was tested.

The many accident case studies in Part II of this book will develop still further the important characteristics of and distinctions between active and latent errors. In addition the book will examine how these may best be detected and corrected before they lead to potentially serious consequences. The two most important forms of latent error, those occurring during maintenance operations and those taking place in the context of management decision-making are discussed in more detail in the following sections.

3.2.4 Latent errors in maintenance

Many serious accidents have resulted from latent errors in maintenance activities although it must be remembered that not all maintenance errors exhibit latency. A relatively simple example of maintenance on an emergency diesel generator demonstrates the effects of latency as well as its potential complexity. If the maintenance task is carried out on components that affect the starting of the engine, and if the error itself is not detected at the time, then it would not be revealed until an attempt was made to start the engine. The error is latent because there will be a delay between the completion of the maintenance task and the requirement to start the engine in an emergency. If the generator provides emergency electrical power to a hazardous process such as a nuclear power plant, then the consequences of such an error could be serious. The standard method of overcoming a latent error in these circumstances would be to initiate a test running of the engine immediately following maintenance.

In the aviation industry, latent errors in maintenance have led to many serious incidents and near misses as well as accidents involving loss of life. An example of such an incident involved a British Aerospace (BAe) 146 aircraft of the Royal Flight following routine engine maintenance in 1997. The incident demonstrates not only latency in maintenance errors but is also an excellent example of how human dependent failure, a subject which is addressed in Chapter 4, can defeat the multiple redundancy provision of having four engines on a jet aircraft. It is therefore an excellent example of how the safety of a highly reliable aircraft such as the BAe146 four engine passenger jet, which is capable of flying on only one engine, can be compromised. The incident, which is described in more detail in Chapter 9, involved a technician taking samples of oil from all four engines for spectrometric analysis. The job also included changing the magnetic chip detection plugs in each engine. The latent error that occurred was an omission to fit a set of 'O' ring oil seals between the plug and the engine casing when replacing the plugs in all four engines. The result was that when the aircraft later departed for a training flight the crew noticed that the oil quantity gauges for three of the engines were indicating empty, and these were duly shutdown to prevent further damage. The aircraft was quickly diverted to Stansted Airport where it landed safely using the power from the remaining engine.

The latency of this error lies in the fact that the loss of oil did not occur until the engine had been running and the oil systems had been pressurized for some time after the aircraft had taken off. It was perhaps unfortunate that this accident occurred to an aircraft of the Royal Flight although on this occasion no members of the Royal Family were on board. The failure prompted a review of Royal Flight maintenance, which had recently been privatized. The accident occurred principally as result of a violation of procedures due to staff shortages and pressure of work. Violations of procedures are another aspect of human error dealt with later in this chapter. The incident demonstrates how an undetected latent failure can lead to a potentially unrecoverable and dangerous situation in what is essentially an unforgiving system where there was no possibility of recovery once the error has been detected. The example also demonstrates the principle of error dependency (see Section 5.4 of Chapter 5), since having made this particular error on one of the engines, the probability immediately increased that the same error would be made on the other engines with little or no chance of error recovery (also see Chapter 5) after the first error was made.

3.2.5 Latent errors in management

3.2.5.1 *Introduction*

Most management errors are latent in their effects. There are two types of management errors to be considered. First of all there are the usual forms of human error associated with work carried out by managers including errors of commission, omission and substitution in, for instance, selection of data and information, preparing reports and carrying out calculations. Managers are prone to these errors in exactly the same way as those carrying out manual tasks in the workplace. However, these management errors are much less likely to be active in nature since it is rare that an accident would occur immediately after the error is made.

The other category of management error takes place during the decision-making process and is crucial to the study of accident causation as the case studies in Chapter 7 will illustrate. Decision-making errors are always latent and their effects can be delayed for extremely long periods of time, before the errors are detected. They are usually not detected at all until an accident has occurred. Another problem is the difficulty in predicting and understanding how decision-making errors are propagated down through an organization to, eventually, lead to an accident. The main reason for this difficulty is that the activities of management have a very broad effect upon a large number of diverse operations where an accident could occur. The scope for a management decision to cause an accident will be a function of the level in the organization where the error was made; the higher the level the broader the potential effect. It is even possible for the cause of an accident to be traced back to the culture of an organization rather than to a specific management activity. The influence of organizational culture upon the development of accidents is even more complex than the effects of the management decision-making process. Two examples of each type of management error, decision-making and cultural factors are, given below.

3.2.5.2 *Errors in management decision-making*

Failures in decision-making have been identified as a principal cause of a number of major accidents. The explosion at the Nypro chemical works at Flixborough in June 1974 (Parker, 1975) is described in more detail in Chapter 7 of this book. It occurred at a major chemical plant producing nylon monomer by means of the oxidation of cyclohexane, a highly flammable and volatile hydrocarbon, in a chain of reactors. When a split developed in one of the reactors, a decision was made by management to replace the reactor with a temporary 20 inch diameter bypass pipe connected to the adjoining reactors by expansion bellows and inadequately supported by temporary scaffolding. The arrangement was subsequently found to be vulnerable to unforeseen mechanical stresses when operating at high pressure or subject to mechanical impact due to the propagation through the system of slugs of liquid reactant. The result was a failure of the bypass pipe some 3 months after it was installed leading to an immense aerial explosion, the loss of 28 lives and widespread damage to the factory and surrounding villages. The decision to install the bypass was driven principally by commercial considerations without full attention being paid to the safety implications. The fact that such a decision could be made was exacerbated by the absence of a qualified mechanical engineer at the factory who might have vetoed the unsatisfactory design of the bypass. The management decision was strongly influenced by the financial implications of the reactor failure. The correct decision would have been to close down the plant and replace the damaged reactor. However, commercial reality is never quite so simple, since the resulting loss of production would have caused the company to become bankrupted.

The example strongly emphasizes the need to consider the complete circumstances of a management decision-making failure. In the chemical industry, and in any other industry, it would not be unusual to deal with such an emergency using a temporary (if unsatisfactory) solution. In this case, a solution that avoided unacceptable financial consequences for the company, took precedence over safety. In the public inquiry report into the disaster, it was recommended that critical management decisions should always be made 'under conditions of minimum conflicting priorities' (Parker, 1975: p. 34).

When examined in more detail, the failure of management involved not so much the decision to install a bypass as a failure properly to design that bypass for the plant operating conditions. Thus the failure was due in part to the fact that mechanical engineering expertise was not applied to the decision-making process. Whether the same decision would still have been made in the presence of such expertise can never of course be determined. Knowledge that became available during the subsequent public inquiry into the accident would not necessarily have been readily available at the time, and if it had, would not necessarily have been applied. The decision-making error was latent in the sense that the failure of the bypass took place some 3 months after the decision was made to install it. The accident occurred during a start-up of the plant under operating conditions which had most probably not existed before. However this accident was a result of a simple management decision-making failure and it was stated in the public inquiry report that the culture of the Nypro organization was not in question and in fact management was 'conscious of its responsibilities relative to safety'. The causes of other major accidents have, however, been attributed more directly to failures in organizational culture which in turn have led to failures in decision-making. An example of this type of accident is given below.

3.2.5.3 *Failures in organizational culture*
The sinking of the cross-Channel ferry 'Herald of Free Enterprise', also described in more detail in Chapter 7, is an example of an accident whose root causes can clearly be traced back to inadequacies in the management culture of the operating company, Townsend Thoreson. The accident occurred on 16 March 1987 when the ferry capsized a few minutes after leaving the Belgian port of Zeebrugge on what should have been a routine crossing to Dover. The sinking occurred as a result of a failure to close the bow doors to the car deck prior to the vessel leaving port. Ostensibly this was due to the absence of the assistant bosun from his post at the time of departure and a failure of the vessel master to check that the bow doors had been closed. Although some blame was attached to those responsible for operating the ship, the Sheen inquiry (Justice Sheen,

1987) into the disaster concentrated its attention on the many inadequacies in the culture and management systems of Townsend Thoreson, the vessel's owners. Criticisms that arose related to many aspects of the company's operation and culture, none of which alone were the direct cause of the accident, but all of which were important contributors. The effects of these broad cultural failures filtered down and affected a number of aspects of the company's operations, but unfortunately were only revealed when an accident occurred with major fatalities. The aspects of company operations affected by the poor culture which were mentioned in the inquiry included:

- overloading of ferries,
- inadequate systems for checking vessel trim,
- poor communications including not responding to safety concerns of the vessel masters,
- no proper system to account for the number of passengers boarded.

One of the broader cultural failures was that nobody in the company was actually responsible for safety; none of the directors was involved in this area of responsibility. In spite of this, the company stated that safety was 'top of the agenda'. The company claimed that the accident had been caused by a single human error committed by the assistant bosun and that such errors were inevitable in ferry operations. The inquiry did not accept this explanation but pointed to the organizational culture of the company in being crucial to the way employees carried out their work.

The lead that must be taken to develop such a culture must come from the top of the company, beginning at director level and be propagated downwards, with suitable checks being made at the highest level to ensure that this is happening in practice. There was no evidence that this sort of commitment existed at senior levels in the organization. Companies lacking this sort of commitment are easily tempted to shift the blame for an accident to human error at the operational level in order to divert attention from organizational deficiencies. However, accident causation is rarely so simple. Even if it were possible for a single error to cause an accident of this magnitude, then the organizational systems that allowed the error to take place would still clearly be at fault.

3.3 Violations

3.3.1 Introduction

A violation is defined as 'an intended action that has taken place in breach of a set of rules, whether or not these rules are written down, have been developed as part of custom and practice or are implicit within the situation itself' (see Section 3.3.2). A common response to violations by management is to identify the culprit(s), allocate blame and impose sanctions to discourage future violations. There may also be a tendency to punish errors in the same way. However, it has already been shown that most errors are systemic in nature and that before blame (if any) is attributed, the causes of the error need to be more thoroughly investigated. If the causes are not corrected then the error will certainly recur. The discussion below will show that most violations also have an underlying cause. In a similar way to errors, unless this cause is properly addressed it is probable that future violations will occur. The most important task of management is therefore to find out the reason why a violation has occurred. Sometimes it is discovered that the rule which has been violated is no longer applicable to the situation for which it was originally devised. Alternatively the rule may be unworkable or difficult to implement for various reasons associated with the situation. Workers may then be motivated to circumvent the rules to carry out the actions. There is invariably a reason for a violation, and whether or not this is a good reason needs to be investigated before disciplinary action (if appropriate) is instigated. The distinction between errors and violations is discussed below. Chapter 12 describes a number of case studies of accidents which have been caused by violations, perhaps the most notable and serious accident being that of the explosion which took place at the nuclear power plant at Chernobyl in the Soviet Union in 1986.

3.3.2 Definition of a violation

Violations have been defined as being *'deliberate ... deviations from those practices deemed necessary ... to maintain the safe operation of a potentially hazardous system'* (Reason, 1990: p. 195). Reason places violations in a similar category to 'mistakes', with the implicit assumption that violations are intended actions (since mistakes are intended plans that may be carried out correctly but inappropriately, see Section 2.4.3, of Chapter 2). However, Reason also allows for the possibility of erroneous or unintended actions. Importantly, Reason makes it clear that it is only possible to describe violations *'in relation to the social context in which behaviour is governed by operating procedures, codes of practice, rules, etc.'.* This is extremely important in understanding why specific violations have occurred, their root causes and how they can best be prevented in future.

An alternative definition given by a study group on human factors (Advisory Committee on the Safety of Nuclear Installations, 1991) describes violations as errors in which *'the human deliberately contravenes known and established safety rules'.* Implicit in this definition is that all violations are intended but it also refers specifically to 'safety rules'. Depending on how safety rules are defined, this rather limits the scope of the definition to violation of operational rules with safety implications which have not been foreseen by the person violating them.

Violations have also been defined as 'intentional errors' (Gertman *et al.*, 1992), although some may regard this as a contradiction in terms and it does not sit well with the definition of human error adopted in this book as being an inadvertent act. This is further discussed in Section 3.3.3.

A broader definition of a violation, taken within the context of the case studies and material presented in this book, is therefore suggested as:

> *an intended action that has taken place in breach of a set of rules, whether or not these rules be written down, are implicit within the action, or have been developed as part of custom and practice.*

This is broader in the sense that it acknowledges the necessity to interpret what is meant by a 'rule'. Not all activity is governed by identifiable rules, and even if it is still governed, the rules may not always be written down. This broad definition allows actions which are knowingly carried out, for instance, in defiance of best practice and which lead to an undesirable event, to be categorized as violations for the purposes of human error analysis. This theme is pursued further in Section 3.3.4.

3.3.3 Distinction between errors and violations

The principal distinction between an error and a violation, consistent with the broader definition above, lies in the matter of intention and knowledge. By an earlier definition, an error is always an unintended or unknowing act or omission with the potential for an undesired consequence. By contrast, a violation must always meet two conditions:

1. There was some level of intention in violating the rule.
2. There was prior knowledge of the rule being violated (although not necessarily full knowledge of its consequence).

If it is not known that a rule is being violated, then the act must be classed as an error (although in legal terms of course, ignorance of the law is no defence). It follows that, although there *is* such a thing as an unintended violation (an act carried out in ignorance of a rule), for the purposes of this book it will always be treated as an error and not a violation. Thus it is possible for an error to be committed which is also a violation, but a violation cannot be an error because of the intention which is present.

It is important to distinguish between errors and violations because:

(a) although both errors and violations may be systemic in nature, the causes may be entirely different, requiring different corrective actions to prevent a recurrence;
(b) if an act of omission is an error then it may indicate an inadequate system (one which is inducing errors), while if it is a violation, it may indicate a faulty rule (one that is inappropriate or difficult to implement);
(c) the presence or absence of intent is one of the factors which will clearly influence blame attribution (although this book is not principally concerned with the legalistic aspects of blame, this is a feature of a number of major accident inquiries dealt with in case studies in Part II of this book).

While a violation is characterized by intent, the fact remains that to establish intent may in some cases not be easy. A number of case studies later in this book are concerned with design errors, and it is possible that what were understood to be errors may in fact have been violations. This seems to be particularly the case with latent errors, where a considerable time has elapsed between the error and the consequence. By the time an accident occurs, there may be a lot of interest in whether an act was an error or a violation, but not enough information available about the circumstances under which the error/violation was made to establish the truth. This may be due to the lapse of time or the natural tendency to plead error rather than violation, in order to reduce blame.

One of the case studies in Chapter 8, the fire and explosion at BP Grangemouth is relevant to this point (see Section 8.2 of Chapter 8). The error was latent and many years elapsed between the original design error and the accident. The question arises why a safety valve protecting a low-pressure separator from over-pressurization was undersized such that it could not cope with a breakthrough of inflammable gas from an upstream vessel operating at a much higher pressure. The written rule required the safety valve to be sized for thermal relief only (to limit the internal pressure build-up when the vessel is engulfed by fire and discharge the vapourized contents to a safe location). With hindsight, it might also have been considered prudent to size the safety valve to discharge a high-pressure gas breakthrough. However, it was not adequately sized nor was there a specific rule requiring this, except the reasonable duty of care expected of a design engineer. Whether a 'duty of care' can be considered a 'rule' is further discussed in Section 3.3.4.3. The result was that when a gas breakthrough occurred the vessel was pressurized beyond its design limit and exploded causing a fatality. Whether or not this was a violation cannot be known with certainty because it is not known whether the conditions for a violation given above existed (in this case, prior knowledge of a rule). The undersizing of the safety valve did not violate any written rules, but it can be argued that it violated good practice. If the designer was not aware that this was good practice, then an error was made and not a violation. The cause of the accident in the case study in Chapter 8 is therefore assumed to be an error.

3.3.4 Classification of violations

3.3.4.1 *Frequency based classification*

Two categories of violation based on frequency of occurrence have been proposed (Reason, 1990). These are as follows:

- 'Exceptional violations' which take place on an infrequent or *ad hoc* basis and often occur in situations which are unusual and where it may be thought that the normal rules may not apply and can therefore be violated.
- 'Routine violations' which take place on an habitual basis, probably not noticed and therefore uncorrected by management. It is often found that routine violations are of rules that are known to apply but where there is a belief that they are flawed, inappropriate or difficult to implement. The belief may of course be justified.

The situation can exist whereby routine violations start to take on the nature of rules, and the behaviour associated with them begins to resemble rule based behaviour. If the violation of an outdated or inappropriate rule becomes routine, then perversely it may be more 'correct' to violate the rule than to observe it, leading to the curious and con-tradictory situation of a 'violation error'. However, this rather complex and devious logic is not intended to lead to a technical definition and there is very little value in pursuing it further.

3.3.4.2 *Causation based classification*

This classification is suggested in a guide to managers on how to reduce the incidence of industrial violations (Mason *et al.*, 1995) prepared by the human factors in reliability group (HFRG). Two additional categories of violation (in addition to the two frequency categories described above) are proposed based on different causes of the violations:

- 'Situational violations' occur 'because of factors dictated by the employee's immediate workspace environment ... including design and condition of the work area, time pressure, number of staff, supervision, equipment availability and design, and factors outside the organization's control such as weather or time of day'.
- 'Optimizing violations' occur due to a perceived motive to optimize a work situation for a number of possible reasons including boredom, curiosity or a need for excitement (by risk taking). Another perverse reason for violations which is not unknown, is that rules are violated because they are 'made to be broken', although it is not intended for this to be pursued further.

These categories do not really form part of a unified taxonomy along with routine and occasional violations since the latter are based on frequency rather than causation. Situational and optimizing violations could also be exceptional or routine depending on the circumstances.

3.3.4.3 *Skill, rule and knowledge classification of violations*

This classification is based on the skill, rule and knowledge (SRK) taxonomy for human error which is discussed in Section 2.3 of Chapter 2. Within this classification it is possible to decouple the categories of routine and exceptional violations from their frequency basis and place them within the R and K based categories respectively (Collier and Whittingham, 1993). This classification can also be extended to include the category of skill based activity. A proposed taxonomy for these categories is set out briefly below.

1. *Knowledge based violations.* Violations which are exceptional, as defined in Section 3.3.4.1 above, tend to occur in unusual situations which are of low frequency and require some action to be carried out in response to an incident. Such situations may demand knowledge based behaviour. This behaviour, as discussed in Section 2.3.1.3, is used for resolving novel situations for which no pre-existing rules are available and where it may be necessary to resort to first principles. However, an intrinsic difficulty occurs here, because by definition it seems impossible to have a violation in a situation where there are no rules to violate. However, this difficulty is overcome by taking a higher level view of the concept of 'rules', consistent with the broad definition of a violation adopted above (an intended action that has taken place in breach of a set of rules, whether or not these rules are written down, have been developed as part of custom and practice or are implicit within the situation itself).

   In the case of knowledge based violations the rules are implicit within the situation itself. They are not external rules, they are more akin to 'rules of thumb' or rules that are generally accepted as being reasonable. In fact this type of 'rule' can be further refined to include 'the action that a reasonable or experienced person might be expected to carry out in response to a situation'.

   In some situations the correct action might be 'no action' as opposed to an action which could exacerbate the situation or an action carried out with an unpredictable but potentially dangerous outcome. The 'reasonable' approach here would be to avoid precipitate action, or possibly call for expert assistance if the situation allowed that.

Risk taking behaviour, given knowledge of the possible consequences, may well be classed as a knowledge based violation. It is akin to a violation of the 'knowledge' that might or should be applied to the situation by a reasonable or experienced person.

Example: A recent high profile example of a knowledge based violation occurred prior to the serious rail accident at Great Heck, near Selby in Yorkshire (Health and Safety Executive, 2001). The driver of a Land-Rover vehicle and trailer, set out on a long road journey after not having slept the previous night. While travelling along the M62 motorway, he fell asleep at the wheel and his vehicle left the carriageway and continued down a steep road embankment before plunging on to a mainline railway track below. Although the driver of the vehicle was able to escape from the cab, the wreckage had fallen in front of a southbound high-speed passenger train which arrived 60 seconds later, colliding with the vehicle and became derailed. An oncoming northbound freight train then ploughed through the wreckage of the derailed train. The accident resulted in the deaths of ten people on the passenger train.

The driver of the vehicle was later prosecuted on ten counts of causing *death by dangerous driving* (a category of offence which had replaced *death by reckless driving* in 1991). He was sentenced to 5 years in prison. The vehicle was found to be in good working order according to the requirements of the Road Traffic Acts, so no rules were violated in this respect. In addition, since the driver was not the holder of a commercial road licence, no specific safety rule was broken in respect of the amount of rest period which must be taken by a driver before commencing a journey or since a previous journey (UK regulations specify that a commercial driver must have a minimum daily rest of eleven consecutive hours, with a number of caveats, although the amount of sleep period is not specified).

The case against the driver of the vehicle was brought under regulations about 'causing death by dangerous driving'. This offence is defined as one where the standard of driving falls far below what would be expected of a competent and careful driver, and it would be obvious to a competent and careful driver that driving in that way would be dangerous. A lesser offence, careless driving, occurs where a person drives a vehicle without due care and attention, or without reasonable consideration for others. A reckless action such as led to the accident at Great Heck therefore meets the criteria for a knowledge based violation. Although there is a regulation defining 'dangerous driving', the interpretation of that regulation by the courts in specific cases is governed by circumstances implicit within the situation itself in accordance with the above definition of a knowledge based violation.

2. *Rule based violations*. This category is more in accordance with what most people would understand to be a violation, the intentional disregard of a written rule. Such violations are more likely to be 'routine' than 'exceptional' in frequency, although the latter cannot be ruled out. The 'rules' in this category also tend to be 'black and white' rather than a matter of interpretation as with the knowledge based violation defined above. However, regarding enforcement, it is possible that the 'rules' may take the form of 'guidelines' not carrying the same force as rules and

regulations. However, even guidelines are likely to be set out in black and white terms, although they may need interpretation regarding the situations to which they are applicable.

Example: Exceeding the vehicular speed limit on a stretch of road is an excellent example of a rule based violation and has all the attributes of a 'black and white' regulation – either a speed limit is exceeded or it is not, although the penalty may vary according to the amount by which it is exceeded. Exceeding a speed limit can become a habitual or routine violation on empty stretches of road where the excessive speed is unlikely to be detected by a police presence or by roadside cameras. There are many other examples in the area of road safety regulation. In a similar way, the disregard of health and safety regulations, such as not wearing ear protection in noisy areas, would be defined as a rule based violation, since the requirement is based on a measurement of noise levels above which protection must be worn.

3. *Skill based violations.* In one sense these are more closely related to knowledge based than rule based violations. Within the definition of violations adopted above, they equate to 'rules which have developed as part of custom and practice'. They would tend not be written down, but if they were, they would not be regarded as hard and fast rules, since they may not always apply. This would occur in situations where work is normally carried out according to a generally understood and accepted method. However, skill based activity, unlike knowledge based activity, does not by nature demand a great deal of original thinking, reasoning or interpretation. The 'rules' are easily understood and generally well known.

Example: A typical example of a skill based violation occurs in control rooms (for instance in a nuclear or chemical plant) where multiple alarm systems are provided to indicate when process variables have exceeded set limits. When an alarm has been triggered, the operator is able to press a cancel button to silence the alarm even though the variable still exceeds the limits. This enables the discrimination of subsequent audible alarms. If the first alarm had not been cancelled subsequent alarms would be subsumed into the first alarm and not detected. The possible cancellation of an alarm also removes the nuisance and distraction of a continuous alarm sound. Once an alarm has been cancelled, it is assumed that the operator will remember that a variable has been exceeded and that he will take appropriate action. However, it is not unusual in control room situations for the audible alarm on a particular variable to sound repetitively when the variable is fluctuating around the alarm setting. The cancellation of the alarm by the operator then becomes an automatic response, even to the point where the operator may not notice if a second audible alarm on a different variable is initiated. He is therefore likely to cancel the subsequent alarm without checking, assuming it to be the first alarm. This is clearly a violation of a rule that alarms are supposed to be investigated, and good practice would require an operator to do this. It may even be a written rule. However, the automatic nature of the violation is such (requiring no reference to rules as in a rule based violation, nor reference to first principles as in a knowledge based violation) that it would be classed as a skill based violation.

The reality of such a situation is that where the operator is extremely busy, it may be impossible for the operator to check the variable for each individual alarm that sounds, while carrying out his other duties. The problem then lies with the system rather than the operator, both in terms of alarm management (systems are available to prioritize alarms) and provision of adequate manpower.

A similar problem occurs with the automatic warning system (AWS) on trains in the UK (refer to the detailed description in Chapter 10). When a train is following one block behind the train in front, the frequency of AWS alarm cancellations will become extremely high and the operation will become an automatic response. In Chapter 10, the errors that lead to signals passed at danger (SPADs), are regarded as active errors rather than violations, since the inadequacy of the AWS is just one contributing factor. There are many other examples of skill based violations.

In Chapter 2 it was described how human behaviour can move from knowledge based to rule based and from rule based to skill based, as experience is gained in performing a task. The transfer of behaviour type can also occur as a result of limited attentional resources due to high cognitive demand, such that a form of internal prioritization occurs. For instance, tasks which were previously rule based, may be relegated to skill based requiring less resources, so that cognitive resources are made available for a more demanding knowledge based or rule based activity. Similarly, tasks which would previously have been knowledge based may be relegated to rule based behaviour. The same principles can be applied to violations, and that is why the classification of a particular violation as SRK based is not fully deterministic but may have to be changed depending on the situation.

In order to use the SRK classification for violations, it is necessary that they are understood in sufficient detail to be able to place them within the appropriate category. This is mainly because a 'rule' which is violated is not always a simple written rule or instruction. Although this may be a more common understanding of a violation, a broader based definition has purposely been adopted in this book. At the knowledge based end of the spectrum of violations, the rule may be what a reasonable or experienced person might be expected to do. At the skill based end of the spectrum, the rule may just be a commonly accepted practice or rule of thumb. As a result, the distinction between error and violation may be fuzzier in the case of knowledge and skill based categories than in the rule based category. This is because, in the absence of a written rule, it may not always be easy to establish whether there was prior knowledge of the 'rule' which was violated (a necessary condition for a violation, see Section 3.3.3).

3.3.5 The causes and control of violations

3.3.5.1 *Systemic causes*

It is possible to analyse the causes of violations in a similar way to the analysis of the causes of errors, as discussed in Chapter 1. Most violations, as with errors will, when investigated, be found to have a systemic cause. In Section 1.2.1, this was defined as 'an external factor which caused the error or made it more likely'. Systemic causes are therefore related to external or situational factors impacting upon the workplace environment and influencing the way work is carried out. They are factors which would not necessarily cause a violation, but which would, often in association with other factors, influence the likelihood of a violation. They are very similar if not indistinguishable from the performance shaping factors that influence the probability of a human error and which are discussed further in Chapter 4.

The systemic cause will often be an inadequacy of the system intended to support a human task but in fact which fails to do so. Systemic errors and systemic violations are similar in the sense that if the cause is not identified and corrected, the error or violation will recur. However, since violations are by definition intentional, there would not appear to be a class of 'random violations' corresponding to 'random errors' (errors that takes place without any apparent intention or cause) as discussed in Section 1.2. It may however be argued that there are some violations which take place without an identifiable cause, due for example to the perversity of human nature in sometimes desiring to breach a rule for its own sake. It may of course follow from this, that there exists a system inadequacy that allows this to happen.

3.3.5.1.1 Motivation
Due to the intentional nature of violations, it is important to consider not only the systemic causes as discussed above but also, for each systemic cause, the possible motivation to violate the rule. A systemic cause may exist which would make a violation more likely but not inevitable so long as the worker was sufficiently motivated to obey the rule. Motivation relates to internal or cognitive factors which would influence whether a rule is obeyed. It is in fact the equivalent of the 'genotype' of an error applied to a violation. The violation of the rule then becomes the 'phenotype'. The motivation may be governed by the perceived validity or usefulness of the rule (with the possibility always that this perception may be correct and the rule is not in fact valid or useful). Given an identifiable cause of a violation (say a system inadequacy), it is thus helpful to describe the motivation(s), which could then lead to the violation. It will usually be possible to apply corrective measures to both systemic causes and motivation once they have been identified. In Table 3.1 some of the more common systemic causes of violations are tabulated and for each of these, a possible range of motivational factors is suggested. Many of the motivational factors may be common to a range of systemic causes.

3.3.5.2 *Control*
As discussed above, once the causes and motivations of violations have been identified, then measures can usually be specified and implemented to reduce the frequency of future occurrences. These measures may be implemented following a violation (which may or may not have led to an accident) as part of the investigative process. Of course it is better if the potential for violations is identified before the violations occur. This may be possible by means of audits, for instance, although it needs to be accepted that this will never be simple because of the range and diversity of violations that are possible.

Control of violations lies within the aegis of management and is within the power of all those who design and enforce the rules given an understanding of why violations occur. Once the causes have been understood then it is usually quite easy to put the

Table 3.1 Causes and controls of violations

Systemic cause	Motivation	Controls
Design of equipment Example: workers may remove machinery guards in order to gain easier access to moving parts for adjustment or removal of debris	Desire to reduce the degree of physical or mental effort required, reducing inconvenience	Redesign of equipment and human machine interface (HMI) to encourage compliance, if necessary automate the tasks
Excessive task demand	Excessive task demand results in the rules being difficult or impossible to implement motivating violations or short cuts	Task analyses of safety critical work should be undertaken to ensure that demand is within reasonable capability
Inadequate information	Hazard resulting from a violation not clearly defined by rules leading to a belief that it is not important	Clear definition of consequences of violating rules through manual, procedures and rules (also see inadequate training below)
Shortage of time	A rule may be circumvented in order to reduce the time needed to carry out a task, due possibly to a desire to undertake the next task, or finish the work early	Ensure that adequate time is available to carry out critical tasks
Inadequate training • Ignorance of rules	Inadequate understanding of the rules and the reason why rules are in place	Rules should be accompanied by an explanation of their purpose and a description of the
• Faulty risk perception	Lack of appreciation of the consequences of breaking rules	consequences of breaking them; it may be useful for employees to sign for critical rules indicating that they have read and understood them (although this is no guarantee they will not be violated later)
Outdated rules	A mismatch between rules and situation due to a change in the situation which the rules have not kept up with, necessitates a violation in order to carry out the task	Periodic review and update of rules and procedures
Inappropriate incentives	Economic or management pressure to increase productivity may lead to circumvention of safe practices and bypassing of safety features to increase personal gain	Avoid using payment by results or time (e.g. piece work) for safety critical tasks where incentives may conflict with safety (instead design incentives and targets which promote safety)
Group membership	Group shared beliefs and social norms promote conformity and affect acceptability of violations which can become normative behaviour	Ensure membership of groups has a positive influence on safety, through safety competitions, statistics, and encouraging safe and compliant behaviour, use of suggestion schemes
	Peer pressure to violate rules if they conflict with unofficial group norms	Channel desires for innovation and change into constructive channels, involve workers in development and review of rules and procedures

necessary controls in place. When compliance with rules is necessary to achieve safe operation, it is a key responsibility of management routinely to audit compliance and ensure that controls have not been bypassed or ignored. Table 3.1 sets out the typical controls or corrective actions that might be put in place for each of the tabulated causes of violations.

An important aspect of controlling violations is that an investigation is carried out of all violations that come to the attention of management. If it appears that management is ignoring violations it may very quickly be perceived that it is in fact condoning violations. The aim of the investigation must be to identify systemic causes and factors which reduce the motivation to obey rules. Controls can then be implemented preferably in consultation with those who are required to carry out the rules. If those who are required to carry out the rules are involved in their formulation, then this will promote a sense of ownership and increase the likelihood that the rules will be obeyed.

Self-monitoring of compliance with rules by the workforce itself is extremely valuable. This should be accompanied by a scheme for reporting of violations and near misses by the workforce, as well as the reporting of difficulties that are being experienced with existing procedures and rules. It will be beneficial for the scheme to be operated under confidentiality and anonymity in order to avoid problems arising from blame. It is particularly important to allow individuals to report back anonymously in situations where group norms predispose towards violations. In order to achieve this it is necessary to build up the sort of organizational culture which accepts feedback on violations without attribution of blame in order to encourage more open and frank reporting.

Chapter 12 of this book deals with case studies of a number of accidents caused by violations.

References

Advisory Committee on the Safety of Nuclear Installations (1991). *ACSNI Study Group on Human Factors. Second Report: Human Reliability Assessment: A Critical Overview*, London: HMSO.

Collier, S.G. and Whittingham, R.B. (1993). Violations: causes and controls. *Proceedings of a Conference on Safety and Well-being at Work: A Human Factors Approach*, Loughborough University of Technology, 1–2 November 1993.

Gertman, D.I. *et al.* (1992). A method for estimating human error probabilities for decision based errors. *Reliability Engineering and System Safety*, **35**, 127–136.

Health and Safety Executive (2001). *Train Collision at Great Heck near Selby 28 February 2001. HSE Interim Report*, Sudbury, Suffolk: HSE books.

Mason, S. *et al.* (1995). *Improving Compliance with Safety Procedures*, Sudbury, Suffolk: HSE Books.

Reason, J. (1990). *Human Error*, Cambridge: Cambridge University Press.

Roger Jocelyn Parker QC (1975). *The Flixborough Disaster. Report of the Court of Inquiry*, London: HMSO.

Sheen (1987). *The Merchant Shipping Act 1984: MV Herald of Free Enterprise. Report of Court No. 8074*, London: HMSO.

4

Human reliability analysis

4.1 Introduction

In the 1960s, advances in technology, particularly in the field of engineering materials and electronics stimulated by the cold war and the space race, led to a major leap forward in reliability of components and systems. This improved equipment reliability was not, unfortunately, matched by a proportionate increase in human reliability. In fact, as systems became more complex, the difficulties of the human operator in managing the new technologies, were exacerbated and the possibilities for system failure due to human error were increased. Due to the growing imbalance between system reliability and human reliability the need arose for methods to assess the frequency or probability of human error in the operation of technical systems. This need was supported by the developing science of ergonomics which attempted to overcome the problem of human error by addressing how the design of the interface between human and machine could take more account of human capabilities and limitations. The overall aim of ergonomics is to maximize human performance in order to reduce the probability and/or consequence of human error. In line with the definition of human error given earlier, the specific objective is to prevent human actions becoming out-of-tolerance in terms of exceeding some limit of acceptability for a desired system function.

There are two basic approaches to the design of equipment from a human error point of view:

1. *The system-centred approach* is where the emphasis is upon the system rather than the human being. In spite of the wealth of information and advice that is available in the field of ergonomics today, many designers are still tempted to adopt this old-fashioned approach. Their natural inclination is to focus upon the system rather than the human being because it is the system that defines the out-of-tolerance limits. It may be argued that this is because designers do not have sufficient knowledge of human capabilities to take them into account in their designs. However, in many cases understanding human capability involves no more than the application of simple

common sense. The end result of a system-centred approach is a design where the human being must adapt to fit the system rather than the system being designed for the human being. Quite often the discovery that the human being cannot easily adapt to the system is not realized until the product is brought into operation by which time it is too late to make the required changes. The result of designing equipment without the application of ergonomic thinking is that when it is brought into operation, even the most highly trained and well-motivated personnel will make systemic errors.

2. *The user-centred approach* ensures that the design of the system is matched as closely as possible to human capabilities and limitations. This does however require an understanding of these capabilities and limitations. In Part II of this book, numerous examples are given of designs that ignored human capabilities. The aircraft accident at Kegworth, UK, described in Section 11.3 of Chapter 11, is a very simple example of an aircraft cockpit design which presented important engine information in a misleading way. It could be argued that the system was adequate until the pilots were faced with an emergency, when rapid decisions had to be taken based on engine information received from the instrument panel. Under conditions of high workload and stress, the cognitive resources of the pilots were overstretched resulting in limitations which restricted their capacity to assimilate external information, as described in Section 2.3.2.1. As the presentation of that information was less than adequate, the probability of an error increased. It was this factor that resulted in the accident at Kegworth.

Unfortunately, the need to adopt a *user-centred approach* is still, even today, not fully understood by many designers and there are numerous examples of complex technological systems that have been designed mainly with system functionality in mind ignoring the capabilities and limitations of the user. Such systems invariably result in degraded levels of human performance with grave consequences for productivity, equipment availability and safety.

4.2 Measuring human reliability

4.2.1 Introduction

The measurement of human reliability is necessary to provide some assurance that complex technology can be operated effectively with a minimum of human error and to ensure that systems will not be maloperated leading to a serious accident. The requirement to measure human reliability in quantified terms arises from the need to make comparisons between the likelihood of error:

(a) in different tasks
(b) in the same task carried out in different ways and under different conditions.

In turn, this allows critical tasks to be identified and measures to improve human reliability to be compared in terms of their cost benefits. The most common unit of measurement to quantify human reliability in a given situation is that of human error probability (HEP). There are numerous approaches which can be used to estimate HEP, and these are often gathered together under the general title of human reliability analysis (HRA). A few of the more well established approaches will be described in this chapter.

4.2.2 Definitions

HEP is defined as the mathematical ratio between the number of errors occurring in a task and the number of tasks carried out where there is an opportunity for error. HEP values are often extremely small numbers and are therefore commonly expressed in the form of an exponent (see Section 4.2.3). HEP is therefore calculated from the following formula:

$$\text{HEP} = \frac{\text{Number of errors occurring in a task}}{\text{Number of opportunities for error}}$$

The number of opportunities for error is frequently the same as the number of times the task is carried out. Thus, for example, if a person is inputting data into a computer via a keyboard and enters 1000 items of information in an hour but makes five mistakes, the average HEP for that period is 5 divided by 1000 equal to a chance of 5 in 1000 or a probability of 0.005 or 0.5 per cent. A number of different ways of expressing this ratio are described below (those readers who are conversant with probability theory and the use of exponents may prefer to skip the following section).

4.2.3 Expressing probability values

Probability is effectively a ratio and therefore has no units. It can however be expressed in a variety of different ways. Using the example above, the calculation of 5 errors divided by 1000 tasks simply results in an average probability for this particular error of 0.005. From this it is seen that the expression of probability will always be a fraction lying between zero and unity. This fraction can also be expressed as a percentage, although this is not usual in reliability work. The average probability of 0.005 could therefore be shortened to 0.5 per cent.

An alternative and much more convenient method of expressing small numbers (and error probabilities are often *extremely* small) is in exponential units. Exponents avoid the need to write down a large number of zeros after a decimal point. This itself reduces the chance of an error in reading or transposing probability values. In exponential terms, a probability of 0.005 can be expressed as 5×10^{-3} which means exactly the same as 5 divided by 1000. The exponent is -3, because 1000 is *one* followed by *three noughts*. The minus sign represents the *division* by 1000. If 5 were multiplied by a 1000 to give 5000 then this would be represented exponentially as 5×10^3. Now the exponent has become $+3$, the plus sign (although not being shown) representing *multiplication* by 1000.

There is an even more convenient way of expressing exponentials to avoid errors in writing out the superscript value (i.e. -3 as in 10^{-3}). In this convention, a value of

1×10^{-3} would be expressed as 1.0E−03. The letter 'E' stands for 'exponent' and 1.0E−03 would represent 1/1000 while 1.0E+03 would represent 1000. Thus 5.0E−03 represents 5×10^{-3} as above. There is an overriding convention that values between 0.1 and unity are given as a decimal fraction, not as an exponent. Thus a probability of 1 in 3 would be given as 0.33 and not generally as 3.3E−01, in exponential units, although mathematically this is also perfectly acceptable. Another convention for the use of exponential units in probability work is that the fractional values are always rounded up and expressed in exponential terms to two decimal places. Most error probability estimations are incapable of achieving better accuracy than this in any case. Thus, the fractional value of 0.004576 would be expressed in exponential terms as 4.58E−03. In this book, probabilities will generally be expressed in both fractional and exponential terms although there may be exceptions to this rule.

4.2.4 Performance shaping factors

Section 4.3 describes how HEP for common tasks may be assessed. Some of these methods depend upon collecting human error data by observing and recording human errors made during a task. The reason for collecting the data is so that it can be applied to the same task carried out at a different time, in a different place by a different person in order to predict the probability of failure of that task. However the transferability of data between tasks is fraught with difficulty. It arises because the circumstances in which the observed task is carried out are likely to be different to the task to which the data is to be applied. Some of these differences will affect the rates of human error which occur. It is therefore important when collecting error data to also record the circumstances under which the task is carried out. The circumstances which can influence human error are often referred to as performance shaping factors (PSFs). However, the process of observing and recording PSFs can also be problematical, since for any task there may be hundreds of different PSFs affecting the performance of the task. For this reason, it is useful to group PSFs of the same type together in generic groups so that they can be more easily identified. Many different schemes for grouping PSFs have been developed, the scheme shown in Table 4.1 being typical.

There are many other PSFs which could be mentioned. However, Table 4.1 encompasses most of the areas to be considered when assessing HEP. The various methods of assessment of human error often provide their own tailor-made systems for assessing PSFs and these will be discussed as they arise.

4.2.5 Task complexity

One of the most obvious factors affecting human reliability is the complexity or difficulty of the task. Clearly a more difficult task will be less reliable than a simple one. There are however two aspects of task complexity which need to be considered. These are:

1. the difficulty of the task,
2. the capability of the person carrying it out.

Table 4.1 Typical PSFs

Factor	Effect on human error rate
Task familiarity	Familiarity with the task being carried out will reduce human error rates at least down to the point where boredom may take over. Methods of improving vigilance may then have to be adopted.
Available time	Complex tasks may require more time to carry out than simple off-the-cuff tasks and if this time is not available, due perhaps to external constraints, then human error rates will increase.
Ergonomics	It is important that the design of the equipment adequately supports the needs of the operator and if it does not then this will cause human error rates to increase.
Fatigue and stress	Depending on the level of stress, the effects can vary from the merely distracting to the totally incapacitating. At the same time there is an optimal level of arousal and stimulation necessary to maintain vigilance (also see Task familiarity above). Fatigue affects the ability to perform tasks accurately and also increases the influence of other PSFs mentioned in this table.
Attentional demands	Human error rates for a single task carried out on its own may increase considerably if other tasks or distractions compete for attention. Human beings are not particularly good at multitasking.
Availability of plans and procedures combined with level of training	Complex tasks require both experience (see below) and information to be completed successfully. Information should be presented in the form of easily understood procedures and plans which in some cases must be memorized depending upon the task complexity.
Operator experience	Human error rates will depend upon whether the person carrying out the task is a novice or an expert, and this must always be taken into account in assessing HEP. To some degree this will depend upon task complexity. Task complexity is discussed in more detail in Section 4.2.5.

Any major imbalance between difficulty and capability will obviously affect whether an error occurs. This has already been discussed in Chapter 1 and shown in Figure 1.1. Thus, complex or difficult tasks need not necessarily give rise to a higher likelihood of error than simple tasks, so long as the person carrying out the task is sufficiently competent or skilled. However, highly complex tasks may also require a high level of additional support from external systems to aid the person carrying out the task. A complex task undertaken by a skilled person who gives it their full attention, properly supported, may be less prone to human error than a simple task that requires little attention, vigilance or skill. This is especially the case where the simple task is carried out at the same time as other tasks or with high levels of distraction.

The work of air-traffic control (ATC) is an example of an extremely complex task requiring a high level of vigilance and a great deal of external support. The error rate for air-traffic controllers is extremely low in spite of the high demands placed upon them and the stress levels which result from this. In order to maintain aircraft separation in crowded skies, the controller needs to maintain a mental model of the air space he is controlling. In ATC jargon, this is known as the 'picture'. The 'picture' is a three-dimensional mental representation of the airspace and aircraft movements for which the controller is responsible. The model will include not only information about the current position of an aircraft but also knowledge of aircraft type, heading, altitude and speed. The controller is not only required to maintain an up-to-date mental model of the scene, but also needs to be in verbal communication with the pilots of the numerous aircraft, always remembering the aircraft with which he is communicating. Clearly, the consequences of an error in this context are extremely severe.

The system must ensure as far as possible that no single error can lead to a mid-air collision without some prior means of detection and intervention and because of the high workload, it is necessary to provide considerable technical support. This takes the form of devices such as aircraft proximity and track deviation warning systems operating independently of the controller and effectively monitoring his performance. In addition supervisory staff and colleagues continuously monitor the work of controllers. Errors are usually picked up well before an infringement occurs and if errors are made, controllers are immediately referred for retraining. In the event of multiple errors within a set time period, the controller would be given other duties. At a time of ever-increasing air-traffic density, the demands upon air-traffic controllers would seem to be pushing human capability to the limits. Research is being undertaken into the possibility of computerized systems taking over most of the air-traffic controllers' task of ensuring aircraft separation in European airspace. The main duty of the controller would then be to monitor the operation of the automatic system. Ultimately, verbal communication with pilots would become unnecessary except in an emergency, since the ATC computer would directly access the aircraft's flight management system in order to control the flight path. This is an interesting example of how to eliminate human error by taking the human out of the loop.

4.3 Human reliability methods

The two main approaches to the quantification of HEP are classified as database methods and expert judgment methods, respectively. Each is discussed below.

4.3.1 Database methods

4.3.1.1 *Human error data collection*

Database methods generally rely upon observation of human tasks in the workplace, or analysis of records of work carried out. Using this method, the number of errors taking place during the performance of a task is noted each time the task is carried out. Dividing the number of errors by the number of tasks performed provides an estimate of HEP as described above. However, since more than one type of error may occur during the performance of a task it is important to note which types of error have occurred.

4.3.1.2 *Example of error data collection*

During a diesel engine maintenance task, an operator is required to screw a filter onto the sump of the engine. The assembly involves ensuring that a rubber O-ring is inserted

Table 4.2 Error rate data for filter assembly task

Observed error or fault condition	Number of filters assembled	Number of errors	HEP
Failure to insert O-ring	3600	2	0.00055
Failure to exert correct torque	3600	6	0.00167
Oil leak due to human error in filter assembly	3600	8	0.00222

in a recess between the metal surfaces to provide a seal before the filter is screwed into position. The filter is then tightened using a torque wrench to ensure that the compression of the O-ring is correct. Any error in this particular task would be discovered by checking whether oil is leaking from the filter joint when the engine is test run. If a leak is discovered, two possible errors may have occurred. Either the O-ring was omitted, or else the filter was under-tightened allowing oil to seep past the seal, or over-tightened resulting in the seal being damaged. These last two errors can be classified under the generic heading of 'failure to exert correct torque'. In order to keep records of human error rates, the cause of each incident of oil leakage would need to be investigated. The data in Table 4.2 would be typical of that collected for such a task.

If it is required to calculate the probability of an oil leak from the sump/filter assembly, then the two probabilities shown above can be added together to give a total probability of failure of 0.00222 or about 1 failure in 450 assembly operations. Clearly from this analysis the dominant human error is a failure to exert the correct torque contributing about 75 per cent of total failure rate. Since there is a cost penalty in having to refit the filter, it would be important to reduce this particular human error. It should also be noted that the fractional HEPs above, since they are very small, are better expressed in exponential terms. Thus the error of 0.00055 would be given as 5.5E−04 and the error of 0.00167 as 1.7E−03. The total error probability is 2.2E−03.

All database approaches to HEP will have been based at some point on the observation of human errors in the workplace or the analysis of records. However, collection of data on human reliability is not common unless a particular problem with human error has occurred. In high-risk industries it may be important to predict the probability of critical human errors, investigate their causes and specify measures for their prevention. The nuclear industry is a good example of a high-risk industry where human error data collection has taken place. Following the Three Mile Island nuclear accident in the US in 1975, many human error data collection programs were commissioned in the nuclear industry. Arising out of this a number of database methods of estimating HEP were developed, mainly involving tasks in nuclear power plant operations. However the same methods were also adopted outside the nuclear industry since there are many similarities between common tasks in nuclear plants, chemical plants, assembly lines and other industrial operations.

Although numerous database methods have been developed only two of the more prominent methods which are in common and accepted use today, will be described here. These are the technique for human error rate prediction (THERP) and the human error assessment and reduction technique (HEART).

4.3.1.3 *Technique for human error rate prediction*

In the 1950s and 1960s, it was found necessary to quantify the potential for human error in order to be able to assess and rank the risks arising from human failure in safety critical operations. The Sandia National Laboratories in the US undertook much of the early work in this field especially in the area of weapons and aerospace systems development. The first industrial HRA studies, however, concerned nuclear power plants. One of the earliest large-scale studies of nuclear safety was the WASH-1400 Reactor Safety Study which commenced in 1972 (Rasmussen, 1975). This used the THERP HRA method to gain an understanding of the impact of human error in a probabilistic assessment of the risk to the public from two nuclear power plants. It estimated the probabilities of accidents involving radioactivity releases including the estimation of health effects, relative to other risks. The study determined that the risk to the public from a nuclear accident was negligible compared with other more common risks.

Development of the THERP method began in 1961 in the US at Sandia National Laboratories and the developed method was finally released for public use in a document NUREG 1278 in 1983 (Swain and Guttmann, 1983). The stated purpose is to present methods, models and estimates of HEPs to enable analysts to make predictions of the occurrence of human errors in nuclear power plant operations, particularly those that affect the availability or reliability of engineered safety systems and components. The *THERP Handbook* assists the user to recognize error-likely equipment design, plant policies, written procedures and other problems in nuclear power plant tasks with a view to reducing the incidence of human error. The method describes in detail all the relevant PSFs which may be encountered and provides methods of estimating their impact on HEP. It also proposes methods of combining the HEPs assessed for individual tasks in the form of a model so that the failure probability for a complete procedure can be calculated. This is carried out by using a method of modelling procedures in the form of HRA event trees. The interaction between individual human errors can then be more easily examined and the contribution of those errors to the overall failure probability of the procedure can be quantified. The HRA event tree is an extremely powerful tool which can be used to analyse safety critical procedures in order to identify the major points of failure. Aspects of human error modelling are further discussed in Chapter 5.

While the THERP methodology was developed within the nuclear industry, it can also provide HEP data for a wide range of industrial tasks found in chemical and other processing industries such as oil and gas production, chemicals and refinery operations. In order to overcome the problem of transferring human error data from one situation to another, the data tables in THERP often provide corrections for common PSFs, including stress factors, the use of written procedures, administrative controls and plant policy. For example, THERP provides a range of HEPs for common errors of commission (ECOMs) and errors of omission (EOMs).

4.3.1.3.1 Errors of commission

Typical data presented by the THERP methodology for estimating the probability of ECOMs in common tasks relates to the operation of manual controls on a power station control panel. One of the more common errors in this type of task is the selection of the wrong control or item of equipment, prior to making an adjustment or change to the process. This error data is later used in an example of human error modelling in Chapter 6. The THERP method presents an error 'Select the wrong circuit breaker in a group of circuit breakers' (*THERP Handbook* Table 13.3). However, the selection error could equally apply to the selection of other controls or items of equipment that are of a similar design and function and are grouped together. The HEP of this and other types of error depends of course upon the local PSFs and THERP invariably includes typical PSFs when presenting the data.

For the selection error 'Select the wrong circuit breaker in a group of circuit breakers', THERP provides two alternative PSFs. For the situation where the circuit breakers (or other controls) are 'densely grouped and identified by label only' THERP suggests an error probability of 0.005 or 5.0E−03 as being applicable with an error factor (EF) of 3. The EF recognizes that the data presented has been averaged over an extremely high number of tasks and is therefore subject to uncertainty. The EF reflects the uncertainty bounds, the lower bound being calculated by dividing the HEP by the EF and the upper bound obtained by multiplying the HEP by the EF. The average HEP for the selection error is 0.005 or 5.0E−03 and the EF is 3. Thus this estimate lies within the possible range of from 0.005/3 = 0.0017 or 1.7E−03 up to 0.005 × 3 = 0.015 or 1.5E−02. There is thus almost an order of magnitude between the upper and lower bounds. The position of the HEP value within the uncertainty bounds depends upon the local PSFs. If the PSFs, that is the grouping and labelling of circuit breakers, is considered average then the value of 5.0E−03 will be used, but adjusted upwards if these factors are less favourable and downwards if they are more favourable.

For the situation where the circuit breakers (or other controls) have 'more favourable PSFs' THERP suggests a lower error probability of 0.003 or 3.0E−03 as being applicable, also with an EF of 3. Thus the HEP now lies within the possible range of from 0.003/3 = 0.001 or 1.0E−03 up to 0.003 × 3 = 0.009 or 9.0E−03. More favourable PSFs might be found if the items are not closely grouped but are more logically grouped to aid recognition. For instance, if the circuit breaker was associated with a current ammeter, which needed to be checked when the breaker was operated then a selection error is less likely, since the ammeter provides a semi-independent check and an alternative method of identification apart from the label.

THERP provides alternative data sets for selection errors (*THERP Handbook* Table 14.1) for locally operated valves where more detailed information about PSFs is available, such as clarity and ambiguity of labelling, size and shape of items which may help to discriminate between items and the presence of unique 'tags' (temporary means of identification of an item of equipment, such as the attachment of a coloured disc or other means of distinguishing one item from a group of similar items). Depending on the PSFs the HEP values then vary from as high as 0.01 or 1 error per 100 selections (unclearly or ambiguously labelled and similar in size, shape and tags) to as low as 0.001 or 1 error in 1000 selections (clear and unambiguous labelling and set apart from other similar items).

Badly grouped engine instrumentation in an aircraft cockpit led to a catastrophic visual selection error and resulted in the Kegworth accident, the most serious air accident in the UK in many decades. This accident and its root cause is described in Chapter 11.

4.3.1.3.2 Errors of omission
The THERP methodology also provides data for estimating the probability of EOMs in common power station tasks. Typical of these and a very common error in many industrial operations is the omission of steps from a written procedure (*THERP Handbook* Table 15.3). THERP distinguishes between procedures comprising less than ten and more than ten steps, the greater the number of steps the higher the probability of an error by, according to THERP, a factor of 3. The best HEP which can be obtained is 0.001 or 1 omission per 1000 steps while using procedures of less than ten steps, but where the procedure is subject to subsequent checking. For more than ten steps this would increase to 0.003. Where there is no checking, the HEP starts at 0.003 for less than ten steps increasing to 0.009 for greater than ten steps. In all cases the EF is 3. Where a procedure should be used but is not used the HEP is given as 0.05 or 1 error in 20 steps with an EF of 5.

4.3.1.3.3 Use of the technique for human error rate prediction human error data tables
The brief selection of data for EOMs and ECOMs presented above is an example of the detail and type of data presented by the THERP method. These particular types have been selected because similar data is used to quantify human errors in a procedure used as an example in Chapter 6. The reader is referred to the THERP method for the provision of error data for other error types and tasks. The usefulness of the THERP data is not so much in the provision of absolute HEP values, but in the logical, comparative and intuitive way in which it is set out enabling rapid comparative estimates to be made of HEPs so that the effect of improvements to PSFs can be more positively assessed. The main disadvantage of the THERP methodology is that the data was collected over 30 years ago from operations using 1960–1970s technology and equipment. It therefore fails to provide failure rate data for operation of the human–computer interfaces in more modern control rooms. The HEART method described below, is however, perfectly capable of filling this gap since it uses a more generic approach to the generation of human error data.

4.3.1.3.4 Additional references
The complete THERP method as developed prior to 1983 by Swain and Guttmann may be found in document NUREG/CR1278 (Swain and Guttmann, 1983). Later and possibly more up-to-date references to this original work are available. Gertman and Blackman (1994) provides a summary of the THERP method and supplies the error data tables. A step-by-step approach to THERP may be found in Dougherty and Fragola (1988).

4.3.1.4 *Human error assessment and reduction technique*
The HEART methodology (Williams, 1988) does not use human error data collected in specific task situations, but rather groups a number of task types in terms of the generic HEP range likely to be achieved. The method uses the concept of error producing conditions (EPCs) defined as factors that can affect human performance making it less reliable than it would otherwise be. Thus, if it is required to estimate the HEP for a particular task, that task must first be placed in one of the main HEART generic groups. These groups are shown in Table 4.3.

Table 4.3 Main HEART generic task groups

ID	Generic task	NHEP	Uncertainty bounds
A	Totally unfamiliar task performed at speed with no real idea of the likely consequences	0.55	0.35–0.97
B	Shift or restore systems to a new or original state at a single attempt without supervision or procedures	0.26	0.14–0.42
C	Complex task requiring a higher level of understanding and skill	0.16	0.12–0.28
D	Fairly simple task performed rapidly or given insufficient or inadequate attention	0.09	0.06–0.13
E	Routine highly practised rapid task involving relatively low level of skill	0.02	0.007–0.045
F	Restore or shift system to an original or new state following procedures with some checking	0.003	0.0008–0.007
G	Completely familiar well-designed, highly practised, routine task occurring several times per hour, performed to highest possible standards by highly motivated, highly trained and experienced persons totally aware of implications of failure, with time to correct potential error but without the benefit of significant job aids	0.0004	0.00008–0.007

Unlike the THERP methodology, HEART does not provide human error data for specific tasks but requires the user to place the task which is being assessed into one of the generic groups in order to establish a nominal human error probability (NHEP). The NHEP is then corrected for the PSFs which are present. In HEART, the PSFs are referred to as EPCs. The range of EPCs provided by HEART are extremely diverse and have been gathered from a wide variety of sources which are referenced by the methodology. A total of thirty-eight EPCs are provided and examples of these are presented in Table 4.4. They are grouped into EPCs having a high, medium and low influence upon the NHEP in order to provide a representative selection. The degree of influence is expressed as a basic correction factor (BCF). This is defined by HEART as 'the maximum predicted nominal amount by which unreliability might change going from "good" conditions to "bad" conditions'.

The BCFs are not applied directly to the NHEP, but by means of a mathematical formula which also includes an assessment of the proportion of affect (PA). The PA represents the degree to which the selected EPC actually has an effect in increasing the HEP for the task. The formula which is used is as follows:

Assessed effect of EPC on error probability = [(BCF − 1) × PA] + 1

This factor is calculated for each EPC that is present in a situation. All the factors are then multiplied together to give an overall assessed effect that is then multiplied by the nominal error probability estimated from Table 4.3. Thus, for instance, if a task fell into generic task group *F* in Table 4.3, the NHEP would be 0.003 and it would be necessary to select the applicable EPCs to correct this probability. Assuming that the EPCs present were Nos 2, 8 and 14, from Table 4.4, then the assessed effect of each would be

Table 4.4 Selection of HEART EPC

No.	EPC	BCF for NHEP
High influence EPCs (BCF 5–20)		
1	Unfamiliarity with a situation which is potentially important but which only occurs infrequently or which is novel	×17
2	Shortage of time available for error detection and correction	×11
3	A low signal noise ratio	×10
4	A mismatch between operators model of the world and that imagined by designer	×8
5	The need to unlearn a technique and apply one which requires the application of an opposing philosophy	×6
6	Ambiguity in the required performance standards	×5
Medium influence EPCs (BCF 2–5)		
7	Poor, ambiguous or ill-matched system feedback	×4
8	Operator inexperience	×3
9	A conflict between immediate and long term objectives	×2.5
10	A mismatch between educational achievement of an individual and the requirements of the task	×2
Low influence EPCs (BCF 1–2)		
11	Unclear allocation of function and responsibility	×1.6
12	No obvious way to keep track of progress during an activity	×1.4
13	High emotional stress	×1.3
14	Low workforce morale	×1.2
15	Disruption of normal work–sleep cycles	×1.1

Table 4.5 Example: Calculation of assessed effect on HEP using HEART

EPC	BCF	Assessed PA	Calculation	Assessed effect
Shortage of time	11	0.3	$[(11 - 1) \times 0.3] + 1$	4.0
Operator inexperience	3	0.1	$[(3 - 1) \times 0.1] + 1$	1.2
Low morale	1.2	0.2	$[(1.2 - 1) \times 0.2] + 1$	1.04

calculated as shown in Table 4.5. Factors for PA have been assumed but are typical of those found in a HEART assessment.

The corrected HEP is therefore calculated from:

$$\text{NHEP} \times \text{EPC2} \times \text{EPC8} \times \text{EPC14} = 0.003 \times 4.0 \times 1.2 \times 1.04 = 0.015$$

It can be seen that the NHEP has been increased by a factor of 5. One of the perceived problems of HEART is that of double counting. If an excessive number of EPCs are identified, some of these possibly having a similar influence on the NHEP, then the multiplicative effect of all these BCFs will produce highly pessimistic estimations of HEP. While in safety and risk calculations it is probably better to err on the pessimistic side, it has been found that HEART can produce unrealistic results if not used by an experienced assessor aware of the pitfalls. Thus while HEART is extremely flexible in terms of the task types it can be used for, overzealous selection of EPCs can lead to unrealistically high estimates of HEP.

On the positive side, HEART has the advantage of being an error reduction as well as an error estimation technique. Once the EPCs have been identified, it is possible to select from a range of error reduction strategies suggested by the method. Hence, for instance, the EPC of time shortage is accompanied by a remedial measure which includes trying to ensure that 'sensitive decisions are not taken against the clock'. All the other EPCs are accompanied by a suggested remedial measure. The HEART methodology has been classed here as a database method, since its generic HEP values are based on data collected from numerous sources for a wide range of human errors which fit into these generic types. Additionally, the EPCs presented in HEART, and their effects upon HEP, have been collated from a wide variety of experimental and observational studies, all of which are referenced (Williams, 1988). A proprietary computer based version of HEART (Williams, 1996) is also available and provides guidance in the selection of generic tasks, EPCs and PA together with warnings on the potential for double counting. There are also recommendations for reducing error potential.

It is obvious from the above example that HEART requires a degree of subjective judgment in the selection of generic probabilities and applicable EPCs in order to arrive at any degree of accuracy in the estimation of HEP. Some methods, however, are almost completely based on subjective assessment, and are referred to as expert judgment techniques. One of these is described in Section 4.3.2 below.

4.3.2 Expert judgment methods

4.3.2.1 *Introduction*

Expert judgment methods of estimating human error are used to overcome either of two common pitfalls:

(a) a shortfall in human error data for the tasks of interest or
(b) a perception that the data which is available cannot reliably be transferred from the situation where it has been gathered into a different situation.

Expert judgment methods rely upon calibration of a psychological estimation scale for human error using numerical probability values by a group of experts, having the ability, experience or knowledge to make valid assessments of human reliability in a given situation. The expertise which is required is of two types:

1. *Substantive expertise* where the expert must understand the situation or task being analysed, which in practice usually amounts to a detailed knowledge of the type of industrial operation being carried out together with the relevant situational factors.
2. *Normative expertise* where the expert is able to translate psychological judgments about the likelihood of an error into numerical values of HEP including the effects of PSFs.

Both types of expertise are rarely present in a single individual, and expert judgment methods generally require the participation of a team of experts having different types of expertise and who are able to reach a consensus through information sharing. Unfortunately, this tends to make the approach rather laborious and expensive and it would only be used to analyse tasks or procedures that were seen to be highly safety critical. Moreover, it has been found that even qualified experts are unable to be entirely objective in their judgments and are subject to various forms of bias arising out of group interaction, often arising out of the need to reach an agreed common estimate. For example, as in any group situation, it is possible that stronger personalities may rule the discussion and override the expert judgment of other individuals who may feel to be under pressure to reach a consensus.

Most expert judgment methods of estimating human reliability have mainly been developed within the nuclear power industry. Very few of these methods are in use today, therefore just one of the more common approaches will be described here, the method referred to as absolute probability judgment (APJ).

4.3.2.2 *Absolute probability judgment method*

In the APJ method a consensus group approach is adopted, each task being analysed in turn, and in isolation, to obtain a consensus estimate of HEP before moving to the next task. It is usual to start with a simple task where it may be possible to obtain actual HEP data for this or a similar task in order to provide an 'anchor point' for the subsequent discussions. This builds both expertise and confidence in the process. It may also be useful for the sessions to be guided by a facilitator who does not provide input to the process in terms of HEP estimation, but is able to prepare and provide information and documentation relating to the tasks to be assessed. Since the facilitator is independent of the process, he may also be able to detect and remedy any bias occurring in the group discussions. Once the first task has been assessed, preferably using an anchor point to determine the initial HEP estimate, other tasks may then be assessed in the same way, using the first result as a benchmark. However as the process proceeds, it is important to carry out a 'sanity check', by reviewing the results objectively and in isolation from the benchmark in order to ensure that a sense of realism is maintained. Additionally, records should be kept of each assessor's subjective estimate prior to a consensus being reached, so that at the end of the process an attempt can be made by the facilitator or others to assess inter-judge consistency. If a consensus cannot be reached in a reasonable time then the individual estimates must be aggregated to produce a result. It is also important to estimate the uncertainty bands applicable to the results.

4.3.3 Conclusion

The main difficulty with expert judgement methods is the need to identify and bring together the required expertise. The methods can therefore be time consuming and resource intensive. In addition it may be difficult to achieve full confidence in the results due to perceived subjectivity and suspected bias. Database methods are therefore probably more acceptable and practicable. The THERP method presents actual data collected in the context of nuclear power generation and provides in-depth guidance on its use together with detailed PSFs. The data are easily transferable to similar tasks within the processing industries. However, it is largely based on 1960–1970s technology and does not include data for tasks associated with more modern control room computer interfaces for instance. Nor would it be possible to use THERP to generate HEP estimates for train-driving tasks for example and similar non-processing industry activities. The HEART method is therefore almost certainly the method of choice where it is judged that the THERP data is not applicable and no other task specific data is available. It is capable of generating probability estimates for almost any industrial task, given that there is sufficient information available about the task to place it accurately within a generic group and realistically identify the EPCs that apply.

All HRA methods have this in common; they require information about the task within which an error of interest may occur, in particular the PSFs which can influence HEP. In order to provide this information it is usually necessary to undertake some form of task decomposition, or breaking down a human activity or procedure into its component tasks presented at a level of detail where it is possible to identify potential errors. In addition, the level of detail must be such that error data is available or can be synthesized for the errors that are identified so that the HEP can be quantified.

4.4 Task decomposition

4.4.1 Introduction

Task decomposition is usually carried out using a technique such as task analysis. The purpose of task analysis is to break down the level of detail of a human activity of interest, such as may be described in a written procedure, to its basic tasks in a systematic way. It is important to decompose the activity down to a level of detail where potential errors present may be more easily identified and even more importantly to a level where data is available for the quantification of these errors. Thus, for instance, there would be no purpose in describing a maintenance activity purely in terms of 'oil filter cleaning' for the purposes of a HRA. In the first place there is no error data available for 'failure to clean an oil filter', nor are any errors associated with this failure actually identified so that HRA can be carried out. It is therefore necessary to break down the separate actions in this procedure using task analysis. The following section provides a brief outline of the technique using a worked example as a demonstration.

4.4.2 Task analysis

4.4.2.1 *Introduction*

Task analysis comprises a formal and systematic means of representing the component tasks which are necessary to implement even the simplest procedures. Apart from being useful in identifying and analysing human errors in action sequences and procedures, task analysis provides a number of other important functions such as those described below (Kirwan and Ainsworth, 1992):

- Allocation of function between man and machine to ensure that the parts of the task which are carried out by a human operator are within his capabilities and that the parts which are outside his capability are, where possible, allocated to a machine which is suitable for the purpose.
- Task specification that defines the work requirements and thus the required capability of the person to be selected and enables suitable training and organization of work to be developed.
- Person specification which enables the right person to be selected to carry out the work.
- Performance assessment enabling, among other things, the identification of potential human errors within a task sequence and in some cases providing input to an assessment of the probability of an error using HRA as described above.

Two main approaches exist to representing a task analysis; these are hierarchical and sequential presentations. The selection of the most suitable approach depends upon the purpose of the task analysis. Each approach is described below using an illustrative example.

4.4.2.2 *Example of a task analysis used for human reliability analysis*

In order to illustrate the use of a task analysis to assist in the preparation of a HRA, an example is provided of an operator procedure for recovering from an exhaust fan failure on a batch reactor plant based on a typical chemical industry process. The process is described in sufficient detail to illustrate the principles involved in identifying and quantifying the potential human errors. The reason for carrying out the task analysis is to investigate the human errors which might occur during operator intervention to prevent a toxic gas released to the building where the reactor is located. The HRA is dealt with in more detail later in this chapter, and the modelling of the operator response is described in Chapter 6 using the same example. Figure 4.1 provides an outline diagram of the batch reactor process (reproduced by kind permission of Electrowatt-Ekono (UK) Ltd).

4.4.2.2.1 Description of process

While this description of the process may seem very detailed it is necessary in order to understand the identification of potential human errors in Section 4.5 and modelling of the errors in Chapter 6 where this example is again used.

The process includes a batch reactor operating over an 8-hour cycle. At the start of the cycle, the reactor is topped up with fresh solvent liquor. The solvent liquor dissolves a crushed solid which is added from a charge hopper intermittently over the 8-hour period. A reaction takes place between the solid and the liquor in which it is dissolved and this gives off a small amount of heat. For the reaction to take place, the contents must be maintained at an elevated temperature just below the boiling point of the reactant mixture (the elevated temperature is necessary to increase the reaction rate and to provide an economic yield while minimizing the formation of unwanted by-products). As a result,

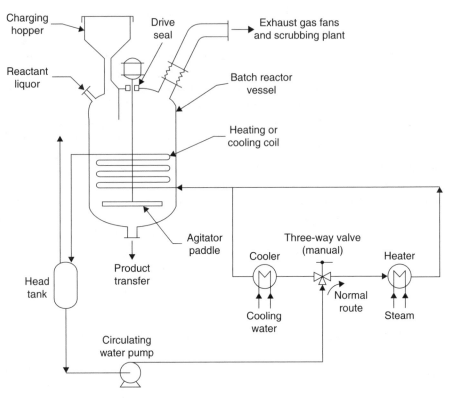

Figure 4.1 Schematic of batch reactor process.

toxic vapours are formed and these must be prevented from entering the process environment. The heat from the reaction is insufficient to maintain this temperature and is augmented by a coil in the reactor through which hot water (heated by steam in a separate heater) is circulated via a pump. An electrically driven agitator paddle is used to improve heat transfer from the coil as well as promoting mixing of the materials in the batch reactor.

The toxic gases, mentioned above, are removed from the reactor by means of an exhaust gas fan. The fan discharges these gases to another plant (not shown) where they are safely removed. In order to prevent the gases entering the building a slight vacuum is maintained in the reactor such that air is drawn into the top of the reactor at a number of ingress points (including the charging hopper discharge valve and the agitator drive seal).

If there is a failure or stoppage of the exhaust gas fans, the vacuum will be lost and there can be a dangerous seepage of toxic gas into the building from the openings in the reactor. If hot water continues to circulate through the coil, the concentration of toxic gas in the building will rise to dangerous levels in about 10 minutes. If the hot water circulating through the coil is stopped then as a result of the heat generated by the reaction, a release of toxic gases will still occur but will take about 40 minutes.

In order completely to rule out the possibility of a toxic gas release to the building, it is necessary for the operator to put cold rather than heated water through the reactor coil to reduce the temperature of the reactor contents. The cooling of the contents is carried out by diverting the circulating water, normally passing through the steam heater, to a cooler fed by cooling water. This is done using a manually operated three-way valve.

In order to prevent a toxic gas release following the failure of the exhaust fan, a reactor shutdown sequence must be put into operation immediately. The failure or stoppage of a fan is indicated by an alarm as well as by a number of other indications that the operator is required to check to assure himself that the alarm signifying a fan failure is not spurious. (This is important because other hazards can arise from a reactor shutdown associated with the need to dispose of an incomplete batch in a safe manner.) These other indications comprise the exhaust fan discharge pressure and the differential pressure across the scrubbing plant after the fan that removes the vapours.

The first action of the operator in the event of an exhaust fan stoppage is to proceed to a plant services room in order to operate the three-way valve to divert cooling water to the reactor. However, before he carries out this action, he should also check the operation of the circulating water pump. If this pump has also stopped, then the supply of cold water will have no effect. He can check at the control panel whether the pump is operating. If it has stopped, then he must restart it in the relevant switch room. If the operator fails to detect that the circulating water pump has stopped and it is not restarted then in about 40 minutes time a major toxic gas release will occur.

It is also possible, as an alternative strategy, that the water circulating pump could be stopped temporarily to remove the heating supply from the coil to allow more time for investigation. If, however, he omitted to restart the pump after having diverted the circulating water to cooling, then a dangerous gas release would occur after about 40 minutes.

4.4.2.3 Hierarchical task analysis

A hierarchical task analysis is a top-down approach that begins with a high-level description of a procedure or sequence of actions. This high-level description is then broken down to a lower level description of its component parts.

Figure 4.2 shows the hierarchical task analysis for the operator response to a failure of the exhaust gas fan on the reactor, all actions having to be carried out correctly in order to prevent a toxic gas release to the building. For the purposes of this example the hierarchy of action is shown at four levels, the amount of detail increasing from top to bottom. At the top, the overall procedure is described. At the lowest level, the description is in terms of individual tasks, shown here as the actions necessary to start the circulating water pump. It is of course possible to break these lowest level tasks down to an even more detailed level of sub-task and so on. However, as described above, there is a logical cut-off point beyond which it would serve no useful purpose to further decompose the tasks at an increasing level of detail. This logical cut-off point is determined by:

- *The information available to develop the task analysis.* This will clearly influence the level of detail that can be shown.
- *The purpose of the task analysis.* Reference to the functions of a task analysis listed in Section 4.4.2.1 show that each function may require a different level of detail to

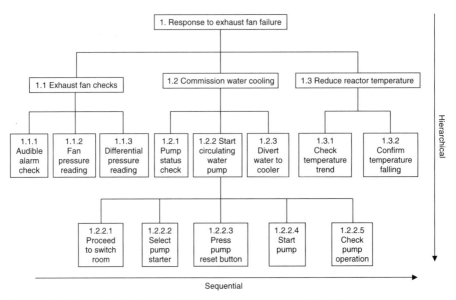

Figure 4.2 Hierarchical task analysis for response to exhaust fan failure.

be shown. For instance, if the task analysis is to be used, as in this example, for identifying potential human errors and these errors are later to be quantified for their probability, then the availability of error probability data will decide the level of detail to be shown.

Hierarchical task analysis is usually set out diagrammatically in the form of a tree, as in Figure 4.2. This type of task analysis is useful for setting out a procedure or plan in terms of overall goals and sub-goals at increasing levels of detail. When the level of detail of the analysis is at the lowest task level in Figure 4.2 (shown only for sub-task 1.2.2) then the hierarchical analysis begins to resemble a detailed sequential task analysis as described below.

4.4.2.4 *Sequential task analysis*

Sequential task analysis is a method of representing a sequence of tasks in the order in which they take place. Any one of the levels of the hierarchical task analysis shown in Figure 4.2 as indicated by the horizontal arrow, could form the basis of a sequential task analysis if the tasks were set out in time order (as they are in this example). However, a sequential analysis would usually be based on the lower, more detailed, levels of the hierarchical analysis since the intention is to provide information about the context and environment in which the tasks are carried out. A sequential analysis would not usually, however, be set out in a diagrammatic way, but in a tabular format, thus allowing a more efficient presentation of information relating to each sub-task in the sequence. An example of the sequential task analysis for the sub-task 1.2.2, the operator starting the circulating water pump, as shown in Figure 4.2, is presented in tabular format in Table 4.6.

Table 4.6 Example of sequential task analysis for the start of the circulating water pump

Main task	Sub-task description	PSF	
		System related	Contextual
1.2.2 Operator starts circulating water pump	1.2.2.1 Operator proceeds to switch room	Switch room is adjacent to control room and cannot be easily mistaken	The sub-task initiates the whole procedure of starting the circulating water pump. In order for it to take place it has already been established that an exhaust gas fan failure has occurred
	1.2.2.2 Operator selects circulating water pump starter	One starter in a set of eight similar starters in the same position. All starters are clearly labelled	A failure here can be recovered if sub-task 1.2.2.5 below is carried out correctly
	1.2.2.3 Operator pushes reset button to reset pump prior to starting	There is only one reset button per starter panel and this is clearly labelled below the button	A failure here can be recovered if sub-task 1.2.2.5 below is carried out correctly
	1.2.2.4 Operator pushes circulating pump start button	Pump starter comprises a spring loaded handle to be moved through 45 degrees, in a clockwise direction to a clearly marked 'start' position	A failure here can be detected by noise and vibration of the contactor solenoids. It can also be recovered if sub-task 1.2.2.5 below is carried out correctly
	1.2.2.5 Operator checks that the circulating pump is running	Since the pump is located at the process, its operation is indicated by the reading of the ammeter of the motor (but not to read the actual value, only the fact, that the needle is displaced). However, it is necessary to select the correct ammeter. Time pressure reduces the incentive to carry out this check	There is no recovery for a failure to check that the pump is running

The information provided about the sub-tasks in Table 4.6 is augmented by a description of the main PSFs which can influence the probability of an error occurring in that sub-task. Two types of PSFs are considered:

1. *System related factors* which are concerned with the working environment and systems provided to support the task, including the design of the man–machine interface and other ergonomic considerations.

    The descriptive notes provided for *system related factors* for the 'start of the circulating water pump' procedure in Table 4.6, mainly concern operational requirements placed upon the operator. These include the need for certain checks and controls, use of hand tools, whether a procedure is required, the quality of labelling of controls and other aspects which can influence whether an error might occur. These, however, are a summary only and in a sequential task analysis it is common to reference other more detailed information in appendices or in separate supporting documentation.

2. *Contextual factors* are concerned with the relationship between the task under analysis and other preceding tasks or task errors which could influence this task as well as the significance of the task for the success of the overall procedure. These factors are examined in more detail in Chapter 5, Human error modelling.

The descriptive notes provided for *contextual factors* for the 'start of the circulating water pump' procedure in Table 4.6 indicate the presence of any interactions with other tasks within the procedure (or even with other procedures) which could influence the probability of an error occurring or in some cases the severity of the consequence of an error that might occur.

The PSFs which have been identified above can be selected from a checklist of typical factors to ensure that all important aspects have systematically been considered and recorded in the task analysis. One of the principal uses of sequential task analyses is for identification of potential errors in a sequence of tasks, and Table 4.6 is extended in Section 4.5 to include a column for potential errors. Methods for identification of errors using task analysis are described below.

4.5 Error identification

4.5.1 Introduction

Task analysis is used to break down the level of detail of a human activity to the point where it is possible to identify human errors for which probability data is available if quantification is to take place. If quantification is not to take place then the process of task analysis provides many qualitative benefits in understanding potential human errors. There are many diverse approaches to identification of errors, and these can usually be categorized as one of two types, taxonomy methods and knowledge based methods. Each is briefly described below.

4.5.2 Taxonomy based methods

Taxonomy based methods use an accepted error classification as the basis for identifying errors; taxonomies such as those described in Chapter 2 would be suitable. Thus, for instance, using the skill, rule or knowledge (SRK) taxonomy, it would be necessary to decide whether a task was SRK based. This would then help to identify how the task might fail. For instance, whereas a skill based or rule based task might fail due to an EOM or ECOM, this is less likely with a knowledge based task, where the errors are more cognitive in nature (tending to be more associated with diagnosis or decision-making). Once a task had been classified as S, R or K based, then the range of error probabilities in Section 2.3 can be used to define the HEP for the task. The SRK classification would not, however, be very helpful in defining the detailed nature of the error and this could hinder an investigation of the cause.

Use of the generic error modelling system (GEMS) taxonomy, on the other hand, would enable a more detailed examination of the types of error applicable to the task. GEMS for instance enables skill based errors to be broken down into slips and lapses. Slips and lapses are effectively ECOMs and EOMs respectively. Rule based errors in GEMS are classified as mistakes or failures in planning, and mistakes are further broken down into rule based mistakes and knowledge based mistakes. However, having defined which of these errors might be applicable to specific tasks, there is probably little data available which is specific to the GEMS approach enabling HEPs to be quantified. Database methods such as THERP will therefore have limited usefulness. The HEART method will be more applicable since it is possible to select generic HEPs by matching the task descriptions in the HEART Generic Task Groups (see Table 4.6 above) to the GEMS error types. However, neither the SRK, nor its derivative, the GEMS approach, is ideally suited to quantitative HRA. These techniques are more appropriate to qualitative assessments of human error potential and for post-accident analysis of events to determine the possible human error causes so that these may be corrected. The knowledge based error identification technique described below is much more useful in identifying errors for quantification.

4.5.3 Knowledge based methods

The knowledge based method uses a checklist of generic error types each of which is tested against the task to ascertain whether there is the potential for an error of that type to cause the task to fail. This enables error types to be matched more precisely and easily to the sub-tasks listed in a procedure such as that described above for the batch reactor process. The description of the generic error type may be further refined by reference, for instance, to the GEMS taxonomy to gain understanding of the causes of the error and enable a more appropriate selection of a HEP. This principle is illustrated by the following checklist of typical generic error types:

- *Error of omission* – this denotes that the appropriate action has not been carried out due to forgetfulness, carelessness, intentional behaviour (due to the violation of a rule) or for no apparent reason at all. However, it is always useful to specify the reason, if known, why the action was not carried out in order to:
  - more precisely select or calculate an appropriate HEP using one of the methods described above;
  - identify methods of preventing a future occurrence of the error.
- *Error of commission* – this refers to the appropriate action not being carried out in the correct manner. Here it is useful to define the reasons why the action might not have been carried out correctly, using for instance, the SRK taxonomy as follows:
  - *Skill based task*: the requirement for the task to be carried out correctly may be defined by the way in which the task is usually carried out, and this in turn can identify incorrect methods of carrying out the task.

- *Rule based task*: the requirement for the task to be carried out correctly is normally defined by a rule or procedure, and knowledge of this will indicate the ways in which the task might have failed.
- *Knowledge based task*: the requirement for the task to be carried out correctly is defined by the action necessary to achieve success. It should be noted that potential knowledge based errors are much more difficult to identify since they usually take the form of planning, decision-making or diagnostic errors where the scope for error is extremely broad.

- *Error of substitution* – this refers to an action being carried out correctly, but in the wrong place during a sequence of actions, possibly because it has been substituted for another task. This also includes the case of a task which is carried out too early or too late, that is, not at the appropriate time in the procedure. A common situation where this can happen is when, following a distraction or interruption to a procedure, the procedure is resumed at a different point, either earlier or later, (but more probably later since an earlier resumption would initiate discovery of the error).
- *Extraneous error* – this refers to the fact that a wrong or non-required action has been performed, whether correctly or incorrectly, in place of an intended action. The situation can also occur following a distraction or interruption to a procedure, when a different procedure is resumed.

These generic categories can be used as a checklist when reviewing a task analysis, such as that set out in Table 4.6, to identify human errors. Table 4.6 is also reproduced as Table 4.7 but with the addition of potential errors based on the generic error types described above. Table 4.6 should be referred to for the PSFs, which have been omitted from Table 4.7.

It should be noted from the above example that for a single sub-task, it is possible to identify more than one potential error that could fail the task. It should also be noted that the potential errors are ordered sequentially. In practice this means using the convention that when a second or subsequent error is identified, there is an implicit assumption that the 'prior error' has not occurred. Thus in sub-task 1.2.2.5 it is assumed that the first part of the task (the decision to check that the pump is running) has been made without omission in order to identify the second identified error (operator checks that *a* circulating pump is running but selects the wrong pump). If the operator had not decided to check any pump (the first error) then the second error could not take place. Thus the human error identification process needs to be ordered in a logical progression if all potential errors are to be revealed.

A similar situation exists with the third error in sub-task 1.2.2.5 (operator checks the right pump but fails to detect it is not running), which can only occur if the first and second errors have not taken place. This convention is used in order to avoid making the assumption explicit in each error description by actually stating that the previous error has not taken place (e.g. the third error would then read 'Having decided to check that the pump is running, and given that the correct pump has been selected, the operator fails to detect it is not running'). This could become rather tedious especially when there are a large number of errors identified sequentially in this way so that the error description could become quite lengthy. The 'prior error' rule is then assumed. The

Table 4.7 Task analysis for the 'start of the circulating water pump' task showing identified errors

Main task	Sub-task description	Potential error	Generic error type
1.2.2 Operator starts circulating water pump	1.2.2.1 Operator proceeds to switch room	Operator proceeds to wrong switch room	Error of substitution
	1.2.2.2 Operator selects circulating water pump starter	Operator selects wrong starter panel	Error of substitution
	1.2.2.3 Operator pushes reset button to reset pump prior to starting	Operator fails to reset pump	Error of omission
	1.2.2.4 Operator pushes circulating pump start button	Operator fails to start pump at all	Error of omission
		Operator fails to start pump successfully	Error of commission
	1.2.2.5 Operator checks that the circulating pump is running	Operator fails to check that the pump is running	Error of omission
		Operator checks that *a* pump is running but selects the wrong pump	Error of commission
		Operator checks the right pump but fails to detect it is not running	Error of commission

convention also promotes more logical thinking in identifying errors making it less likely that errors will not be missed.

The error identification process may in many cases require a more complete understanding of the task than is sometimes provided by the task analysis. In these cases it will be necessary to provide further decomposition of the sub-task. The dependence of errors upon previous actions and whether they were carried out correctly, is described in more detail in Chapter 5 which deals with human error modelling in the context of a complete task or procedure.

References

Dougherty, E.M. and Fragola, J.R. (1988). *Human Reliability Analysis: A Systems Engineering Approach with Nuclear Power Plant Applications*, New York: John Wiley and Sons, Inc.

Gertman, D.I. and Blackman, H.S. (1994). *Human Reliability and Safety Analysis Data Handbook 1st edition*, New York: John Wiley and Sons, Inc.

Kirwan, B. and Ainsworth, L.K. (1992). *A Guide to Task Analysis*, London: Taylor & Francis.

Rasmussen, N. (1975). *Assessment of Accident Risk in US Commercial Nuclear Power Plants*, Washington DC: US Nuclear Regulatory Commission (WASH-1400 NUREG 75/014).

Swain, A.D. and Guttmann, H.E. (1983). *Handbook of Human Reliability Analysis with Emphasis on Nuclear Power Plant Applications. Final Report (NUREG 1278)*, August 1983, Washington DC: US Nuclear Regulatory Commission.

Williams, J.C. (1988). A data-based method for assessing and reducing human error to improve operational performance. In E.W. Hagen (ed.) *Conference Proceedings for 1988 Fourth International Conference on Human Factors and Power Plant*, 5–9 June 1988, Monterey, California, pp. 436–450.

Williams, J.C. (1996). *HEART-PC*, Horsham, UK: Electrowatt-Ekono.

5

Human error modelling

5.1 Introduction

Up until now, errors have been considered as events taking place in isolation although it was seen from the Task Analysis in Chapter 4, that errors within a task can be influenced by other related events taking place in the vicinity or at the same time. This relationship between tasks and error can be explored using techniques of human error modelling. In order to better understand the role of human error in accident sequences as described in the next chapter, it is necessary here to consider the following aspects of human error modelling:

- *error recovery* which, following error detection, allows for its effects to be eliminated or mitigated by a subsequent recovery action
- *error dependency* which takes account of how errors may be made more likely by the occurrence of a previous error.

Each aspect of error modelling is examined in more detail in the following sections. In order adequately to model these aspects, it is necessary to understand simple probability theory. This in turn will enable error probabilities within sequences of activities, such as procedures, to be successfully modelled and compared. Readers familiar with probability theory may wish to skip Section 5.2.

5.2 Basic probability theory

5.2.1 Introduction

It has been shown how human error probability (HEP) may be calculated by dividing the number of errors which occurred during a task by the number of opportunities for error. Thus, if over a period of time one error occurred during thirty identical tasks, then the average probability of that error would be $1/30 = 0.033$ or $3.3E-02$. Since, human errors rarely occur in isolation, but as a part of a sequence of tasks forming an activity, it is necessary to be able to combine error probabilities to produce the joint or combined probability of the failure of the whole activity. This allows the probability of failure of a complete procedure to be calculated from the individual failure probabilities of the errors which could occur within that procedure. In order to combine error probabilities, it is necessary to gain an understanding of the concepts of error dependency and error recovery. It is possible to model these concepts using logic trees such as fault trees or event trees adapted to show the interactions between human errors. The basic concepts of probability theory are set out below.

5.2.2 Failure and success probability

So far, only the probability of a task *not* being carried out correctly, that is, the probability of an error or a failure probability has been considered. It is also possible to express the probability of an error in terms of success probability, or the probability of a task being carried out correctly. Success probability is calculated by subtracting failure probability from unity, thus:

Success probability $= 1 -$ Failure probability

Hence, if the probability of an error is 0.03 or $3.0E-02$ then the probability of an error not occurring, or a task being carried out successfully is $1 - 0.03$, or 0.97. As it lies between 0.1 and unity it is expressed here as a decimal fraction.

5.2.3 Probability of two or more independent events

When the probability of two independent events is known, it is possible to calculate the probability of one or other or both of the events occurring at the same time, if these events are mutually exclusive. Let us assume that two independent events, *a* and *b* each have a probability of occurrence of $P_a = 0.03$ and $P_b = 0.2$, respectively:

1. The addition rule of probabilities states that the probability of *a* OR *b* occurring is equal to the sum of the two independent probabilities minus the product of the two probabilities, thus:

$$\text{Probability of } a \text{ OR } b = P_{(a \text{ or } b)} = (P_a + P_b) - P_a P_b$$
$$= (0.03 + 0.2) - (0.03 \times 0.2)$$
$$P_{(a \text{ or } b)} = 0.23 - 0.006 = 0.224$$

2. The multiplication rule of probabilities states that probability of *a* AND *b* occurring at the same time is equal to the product of the two probabilities, thus:

$$\text{Probability of } a \text{ AND } b = P_{(a \text{ and } b)} = P_a \times P_b = 0.03 \times 0.2$$
$$= 0.006 \text{ or } 6.0E-03$$

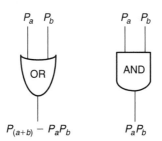

Figure 5.1 Symbolic 'AND' gates and 'OR' gates.

5.2.4 Combining human error probabilities using logic gates

It is possible to express the above concepts in diagrammatic form using logic symbols or 'gates' to represent OR and AND. These symbolic representations can be used in logical trees such as fault trees or event trees to model combinations of human error. There is an advantage in representing human error using logical modelling methods. The ways in which human errors occur and interact with each other can be extremely complex, and modelling these interactions diagrammatically enables a clearer understanding to be gained. At the basic level, human errors can be modelled through OR and AND gates as shown in Figure 5.1.

The left-hand diagram represents the probability of occurrence of events a OR b. The right-hand diagram represents the probability of occurrence of events a AND b simultaneously. The probabilities P_a and P_b are fed through an OR gate and an AND gate, respectively. The outputs from the gates are calculated according to the above formulae. Combinations of AND and OR gates can be used to show the interaction between groups of events using a fault tree or event tree as described below.

5.2.5 Fault trees

The fault tree shown in Figure 5.2 uses the symbolic logic gates described above to combine four events, a AND b, as above, and two new events, c OR d which are combined through an OR gate. The outputs from the AND gate and the OR gate are shown as P_x and P_y and these are combined through another AND gate. We have now combined the probabilities of four events.

Referring to the addition and subtraction rules described above, the combined probability can be calculated as follows, using the probability values in Table 5.1.

The result of the AND gate is $P_x = P_a \times P_b = 0.03 \times 0.2 = 0.006 = 6.0\mathrm{E}{-03}$

The result of the OR gate is $P_y = P_c + P_d - P_{cd} = (0.08 + 0.3) - (0.08 \times 0.3)$
$$= 0.38 - 0.024 = 0.356$$

Therefore the overall
probability is $P_{xy} = P_x \times P_y = 0.006 \times 0.356$
$$= 0.002136$$
$$= 2.14\mathrm{E}{-03}$$

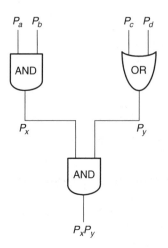

Figure 5.2 Example fault tree.

Table 5.1 Probability values for worked example

Event	Probability of failure
a	0.03 (3.0E−02)
b	0.2
c	0.08 (8.0E−02)
d	0.3

The combined probability of events x and y occurring is therefore 0.00214, which can be expressed in exponential units as 2.14E−03.

The events a, b, c and d in the worked example can represent not only human errors but any sort of failure, including mechanical and electrical components for instance.

5.2.6 Event trees

An alternative approach to modelling a set of events is the event tree. The event tree is a less concise representation than a fault tree, but has the following advantages for modelling human error events:

- The event tree shows more explicitly the various pathways to failure, which can be useful as an aid to understanding how human errors can occur.
- The event tree is able to represent better the sequences of events in time order, including the human errors which can occur during a procedure.

The events represented by the fault tree in Figure 5.2, using the same probability values, can be represented by the event tree shown in Figure 5.3.

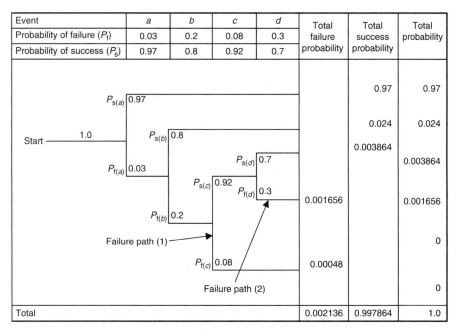

Event	a	b	c	d	Total failure probability	Total success probability	Total probability
Probability of failure (P_f)	0.03	0.2	0.08	0.3			
Probability of success (P_s)	0.97	0.8	0.92	0.7			

Figure 5.3 Event tree representation of Figure 5.2.

The event tree commences with a 'start' probability of 1.0 (i.e. if this was a procedure, the probability of 1.0 means that it is a certainty that the procedure will be commenced). The sequence of events from a to d are shown across the top of the tree, with the probability of failure 'P_f' for each event as shown as in Table 5.1. Events from a to d represent errors occurring in tasks a–d. The tasks from a to d take place in that order. Also shown is the probability of success P_s for each event, derived from the equation as given below:

$$1 - P_s = P_f$$

Each event is represented by a node in the tree, and from each node there proceeds a success branch (upwards and to the right) and a failure branch (downwards and to the right). Each of these branches lead on to another node representing success or failure of the next event in the sequence. Thus moving from left to right, the first node that is reached corresponds to the event a and the probabilities of success and failure, $P_{s(a)}$ and $P_{f(a)}$, respectively, are shown above and below this node according to the values in Table 5.1. According to the fault tree in Figure 5.2 (of which this event tree is an alternative representation) failure of the whole procedure requires both events a and b to fail along with one or other of either event c or d. The event tree therefore represents all the possible failure paths leading to failure of the procedure. In this case there are two possible failure paths:

1. failure of events a and b and c or
2. failure of events a and b and d.

Thus, failure path (1) can be traced via $P_{f(a)}$ through $P_{f(b)}$ and $P_{f(c)}$ to the total failure probability column. Similarly, failure path (2) is traced via $P_{f(a)}$ through $P_{f(b)}$, $P_{s(c)}$ (because in this path event c has not failed but is successful) and $P_{f(d)}$ to the total failure probability column.

The overall path failure probabilities shown in the total failure probability column are calculated by multiplying the failure probabilities out of each node, thus:

Failure path (1) probability = $0.03 \times 0.02 \times 0.08 = 0.00048 = 4.8E{-}04$
Failure path (2) probability = $0.03 \times 0.02 \times 0.92 \times 0.3 = 0.001656 = 1.66E{-}03$

The overall failure probability is calculated from the sum of all the failure paths, and in this case amounts to the total probability of 0.002136, which is the same as the value calculated from the fault tree in Figure 5.2.

In the event tree, the success probabilities have also been calculated in exactly the same way by multiplying together the node probabilities to give the probability of the successful outcome for each of the success paths. These are then summed in the same way to give a total success probability of 0.997864. As expected, the sum of failure and success probabilities (0.002136 + 0.997864) is equal to unity. It can be seen that event trees provide a much more explicit representation of the failure paths than the equivalent fault tree in Figure 5.2. When modelling human error this increases the level of understanding of the interactions between separate human errors in the tree; interactions such as dependency and recovery are described later in this chapter.

HEPs are often combined using a variant of the event tree shown above, known as an human reliability analysis (HRA) event tree. An HRA event tree, for the procedure used as an example in Chapter 4, is given in Chapter 6. The HRA event tree is better able than a fault tree to model successfully the concepts of error recovery and error dependency that are described below.

5.3 Error recovery

5.3.1 Introduction

The role of error recovery in understanding human error cannot be overestimated. If it were not for error recovery, many aspects of our modern lifestyle which we presently take for granted would become chaotic, impossible or extremely hazardous. Driving a car for instance on roads congested with other vehicles and pedestrians would result in mayhem if it were not possible to recover from the frequent driving errors which are made. Car drivers make small errors on a continuous basis in response to a rapid series of external demands made upon their driving skill arising from local traffic and road conditions largely beyond the control of the driver. For instance, small errors in directional control of the car are subject to continuous corrections to the steering by the driver responding to visual feedback, enabling the desired course to be maintained. Likewise, errors in braking, such as late application of the brake pedal, may be corrected as long as a potential collision hazard does not lie within the braking distance of the car. These are rather obvious examples of error recovery. They rely on the person making the error responding to immediate feedback from actions carried out.

In an industrial context, error recovery is equally important. In critical applications it should always be considered whether a single task error could result in an unacceptable consequence and if so it should be ensured that recovery from the error is possible including a reversal of its effects. Recovery is possible if the error is detected and there is sufficient time for recovery before the consequences occur. If recovery or timely detection of the error is not possible and the consequences are unacceptable, the task should be redesigned to eliminate the chance of error. If this is not possible a reliable error recovery mechanism should be engineered into the system.

5.3.2 Error recovery mechanisms

It is important to consider the mechanisms by which errors are recovered. The first fact to note about error recovery is that it does not prevent the error taking place. The recovery applies not to the error but to the potential consequence of the error, which, in favourable circumstances, might be avoided by error recovery. Error recovery cannot occur until the error to be recovered has actually been revealed. If the error is a latent error and is not revealed, then clearly it cannot be recovered at the time. Error recovery is nearly always achieved by means of an independent action following the action where the error first occurred. The recovery action for an error therefore prevents or mitigates the consequence of the error. There are three conditions that must be met for error recovery to occur:

1. *Detection.* The error to be recovered must first have been detected. Latent errors are, by definition, not detected at the time they are made, and therefore there will be some delay in any recovery action. While this gives more time for recovery, latent errors are less likely to be recovered due to the low probability of detection of the error in the first place. Unfortunately, latent errors only tend to be revealed by the consequences by which time the chance of recovery has been lost. Recovery is therefore more likely for active errors, which are detected immediately, than for latent errors.
2. *Recoverability.* The original error must be recoverable. To be recoverable the consequences must be preventable or capable of being mitigated in some way by the recovery action. For this reason many errors are simply unrecoverable. If the consequences of an error are serious, and if that error is likely to take place and is not recoverable, then attention needs to be given to prevention of the error rather than recovery from the error.
3. *Time and opportunity.* There must be sufficient time and/or opportunity available for the error to be recovered. While it may appear that more time is available for recovery from latent errors, because the effects are not immediate, they are less likely to be detected unless a recovery mechanism such as checking or testing, is provided.

Recovery mechanisms generally fall into two categories, planned or unplanned:

- Planned recovery – a recovery mechanism is planned if it takes the form of an engineered barrier against the consequences of the error. Such a barrier would be intentionally designed to detect that an error has been made allowing recovery to take place before the consequences of the error become irreversible. An example of such a mechanism would be using a checklist for a procedure to ensure all tasks had been completed successfully. The automatic warning system (AWS) for UK train drivers described in Chapter 10 is another example.
- Unplanned recovery – a recovery mechanism is unplanned if the recovery results from the accidental discovery that an error has occurred, allowing the consequences to be prevented.

A rather simplistic example of an engineered recovery system comes from the annals of the chemical company ICI Ltd. Many years ago, the company operated a rather dangerous batch reactor under manual temperature control. The reaction taking place was unstable and proceeded explosively if the temperature was allowed to rise too high. If the temperature fell too low, the reaction would not proceed at all. Since this was before the era of automatic temperature control systems, the operator was required to observe the temperature and control it within set limits by manually adjusting the flow of cooling water to a coil inside the reactor. If the temperature rose too high then he would increase the flow of cooling water and *vice versa*. The problem with this task was that although the task itself was undemanding, a high level of vigilance of the operator was required. However, the task was so undemanding it was found that the operator's attention was liable to wander leading to a high frequency of temperature excursions. In spite of the severe consequence of the operator failing to perform his task, it was so boring that even the most diligent of operators would tend to fall asleep. In a modern chemical plant, such a dangerous reactor would be subject to automatic temperature control with independent systems to monitor the temperature and shut the system down if the temperature approached the upper limit. However, at the time this plant was operated, such systems were not employed and an alternative but extremely effective method was provided to protect against high temperature in the reactor due to the loss of vigilance by the operator. The operator was provided with a one-legged stool. If the operator fell asleep he would fall off the stool – an extremely effective way of restoring his attention to the task in hand. This is an early example of an engineered recovery system.

In modern industry more sophisticated error recovery mechanisms will be provided. Some of the best examples can be found in the design of maintenance procedures. A common error in a mechanical assembly process is that of installing components in the wrong order or with the wrong orientation. It is, however, often feasible by the use of carefully designed components to make it impossible for items to be incorrectly installed in this way. An example is the use of locating lugs or interlocking parts that ensure the correct orientation or order of assembly of a component. If an error is made in the assembly of a component, the need to interlock one part with another prevents the installation of that or the next component, thus revealing an error has been made. In assembly line work, the provision of 'mistake proofing' stations, where visual examples of incorrect and correct assembly are displayed, can assist error recovery. The display shows the workers the most common mistakes that are likely to be made with advice on how to avoid them.

5.3.3 Effect of error recovery on error probability

The calculation of error probability is described in Chapter 4 and the way in which error probabilities can be combined is described earlier in this chapter. It is now possible to consider how error recovery can influence the overall probability of an error being made, at least in terms of the consequences being realized. In order to assess this, it is necessary to further examine the earlier definition of a human error. The definition of an error adopted assumes that an undesirable consequence will result. Hence an error resulting in a consequence will always be defined as an un-recovered error. It is possible for this to occur in one of two ways:

1. no recovery action was carried out and a consequence resulted or
2. a recovery action was carried out but failed.

Whenever an error is made which does not lead to a consequence there are also two possibilities:

1. The error is of no significance with regard to the desired outcome, in which case the error is of no interest and can be dismissed.
2. The error was made and no consequence resulted because of a successful recovery action. This is defined as a recovered error.

5.3.3.1 *Probability of human error recovery*

For a human error to occur any potential recovery action must have failed. Assume that P_{he} represents the probability of a human error (and consequence) and that P_r represents the probability of failure of a recovery action. Then from the multiplication rule of probability described above, the overall probability (P_f) that a consequence occurs (i.e. the human error is not recovered) can be calculated from:

$$P_f = P_{he} \times P_r$$

Hence, if the probability of an error P_{he} is 0.03, and the probability of failure of the recovery action P_r is 0.1 (or 1 in 10), then the overall probability of failure is 0.003 or $3.0E-03$.

$$P_f = 0.03 \times 0.1 = 0.003$$

Introducing the possibility of recovery from an error in this case reduces the original error probability by a factor of 10.

Clearly, if the recovery action fails, then the probability of failure of the recovery action is 1.0 and the overall probability of failure P_f and the probability of the original error P_{he} are the same:

$$P_f = 0.03 \times 1.0 = 0.03$$

5.3.3.2 *Recovery factors*

The use of recovery factors provides a practical means of assigning failure probability values for recovery dependent upon the mode of recovery and its reliability. A recovery factor is therefore defined as *any factor which prevents or limits the undesirable consequences of a human error*. Numerically it is the same as P_r, the probability of failure of recovery. One way of determining the recovery factor for a particular situation is to examine a list of potential recovery modes and select those that are present. The greater the number of possible recovery modes, and the more effective each recovery mode is, then the greater is the probability of recovery (and the lower the failure probability P_r). Typical recovery modes have been suggested by (Swain, 1987) in a shortened version of the technique for human error rate prediction (THERP) HRA method (described in Chapter 4) known as the accident sequence evaluation programme (ASEP) developed in the context of nuclear power plant operations. These are summarized in a more generic form below together with the estimated recovery factor P_r:

1. *Compelling feedback*. This comprises annunciators, alarms or other devices built into the system which unambiguously warn an operator that an error has been made. In general this is considered to be an extremely effective recovery mode for active errors. The UK train AWS would fall into this category. Although the system has a number of shortcomings as discussed in Chapter 10, it is still a generally effective means of reminding drivers that they have passed a caution or danger signal and has indubitably prevented many signals being passed at danger. Its presence could probably have prevented the Southall rail accident in 1997 (see Section 10.5.2).

   The recovery factor (the same as P_r) is extremely small and may be assumed for general purposes to be less than 0.0001 (1.0E−04).

2. *Different person check*. This is defined as a check on the results of a task made by a different person to the one who carried out the task ideally using a check list or procedure. The check must be capable of detecting that an error has occurred. It often takes place following maintenance or calibration tasks but also includes operational tasks. It is more effective for recovering latent errors than active errors. It is not as reliable as the compelling feedback recovery mode. It is less susceptible to failure by dependency mechanisms (as discussed in Section 5.4) than the same person check described below.

   The recovery factor or P_r is in the order of 0.001 (1.0E−03).

3. *Same person check*. This is similar to Recovery Mode 2 given above except the check on the results of the task is made by the person who carried out the task. It is assumed a checklist is used. The failure in this check can include either omitting the check altogether, or not carrying it out correctly. In terms of the error of not carrying out the check correctly, it is highly susceptible to failure by a within-person dependency mechanism as discussed in Section 5.4.2. It can recover active or latent errors.

   The recovery factor or P_r is in the order of 0.01 (1.0E−02).

   If this recovery mode and recovery mode 4 below were both present for latent errors, then P_r would fall to 0.001.

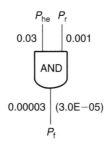

Figure 5.4 AND gate showing error recovery mechanism.

4. *Routine check.* This is without a checklist on an hourly, daily or shift basis to detect faults which may have developed after tasks have been carried out but have not been revealed. It is the least reliable of all the recovery factors.

   The recovery factor or P_r is in the order of 0.1.

The recovery factor is multiplied by the probability of the error without recovery to calculate the overall error probability including recovery. Thus, for example, if the error probability is 0.03 (3.0E−02), and Recovery Mode 3 is present (i.e. a check is carried out by the same person who made the error using a checklist) then the overall error probability including recovery P_f is given by:

$$P_f = P_{he} \times P_r = 0.03 \times 0.01 = 0.0003 = 3.0E{-}04$$

The danger of taking undue credit for recovery factors is that using the multiplication rule of probability, unrealistically low values of HEP can be obtained. This leads to a general principle, used in HEP calculations, of setting a lower limiting value for HEP as described below.

5.3.3.3 *Human error limiting values*

It is possible to use an AND gate to show the effect of error recovery as described in Section 5.3.3.2. This is represented in Figure 5.4.

In this example, Recovery Mode 2 from Section 5.3.3.2 is used, where the recovery factor is 0.001 (1.0E−03). Thus, assuming $P_{he} = 0.03$ then P_f, the overall probability of error including recovery is calculated from:

$$P_f = P_{he} \times P_r = 0.03 \times 0.001 = 0.00003 = 3.0E{-}05$$

using the multiplication rule of probability. A probability of 3.0E−05 represents a HEP of 1 in 300,000. It is extremely low. It is clear from this that using error recovery and the multiplication rule it is possible to obtain extremely small overall HEPs. However, in practice it is extremely difficult if not impossible to obtain HEPs of less than about 1 in 100,000 (1.0E−05). This is effectively a limiting value or lower bound for HEP calculations. The calculated value of P_f from Figure 5.4 is therefore just about possible, but in order to justify this extreme caution would be needed and it would be necessary to ensure that all the recovery modes assumed are indeed realistically achievable in practice.

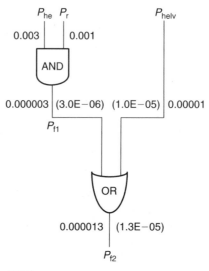

Figure 5.5 Fault tree for HELV.

To show how the lower bound of HEP operates in practice, assume that the probability of human error without recovery, P_{he}, is even lower than in the example above, by say a factor of 10. Thus if P_{he} is 0.003, it can be seen that the calculated value of P_f would be 0.000003 or 3.0E$-$06. Here it would be strongly recommended that a human error limiting value (HELV) be adopted. Although a value of P_f of 0.000003 or 3.0E$-$06 has been calculated, the value to be used should be no greater than 0.00001 or 1.0E$-$05. This operation can also be represented by a fault tree using an OR gate to introduce the HELV. Thus, the fault tree would then be as shown as in Figure 5.5. P_{helv} represents the HELV of 1.0E$-$05.

As P_{f1}, the results of the simple AND gate, is 3.0E$-$06 and unrealistically low, it needs to be modified by the HELV value, $P_{helv} = 1.0E-05$. This is brought in via an OR gate. This logical expression means that P_{f2}, the overall probability of an error including recovery, can occur either via the error and its recovery action (shown by the AND gate) giving a very low probability OR by means of a hypothetical event, the HELV, which has probability of failure of 1.0E$-$05 as decided above. Quite often the hypothetical event can represent a common mode of failure, which will intervene to degrade the recovery action, but which may not necessarily have been identified. The resulting probability, P_{f2}, including a limiting value is then, using the addition rule, the sum of P_{f1} and P_{helv} giving a final value of 0.000013 or 1.3E$-$05. It should be borne in mind that this is merely a logical device to ensure that a lower bound, or HELV, is applied to correct HEPs, which would otherwise be unachievably low.

5.3.3.4 *The use of limiting values in modelling redundancy*
The device to include a lower limiting value has been shown here in the context of error recovery. It is also useful for any situation where the multiplication rule of probabilities is applied to two supposedly independent events and results in an unrealistically low

joint probability for the combined events. For instance, the situation frequently arises where it is intended to claim some form of redundancy for multiple actions or items of equipment. A hazardous process may be provided with two or even three separate safety devices any one of which will ensure that the plant is shut down to a safe condition, even if one of the devices does not operate. Mathematically, assuming the devices are completely independent, the probability of failure of all the devices at the same time is equal to the product of the three separate probabilities, which could result in an extremely low probability for the system failure as a whole.

As an example, if the probability of failure of each of three devices was 0.005 or 5.0E−03, then using the probability rule, the probability of failure of all three devices at the same time would be about $(0.005)^3$ or 1.25E−07. This is equivalent to one failure in 8 million demands, which may not be achievable in practice. Some other common cause failure which would affect all three safety devices at the same time would almost certainly intervene at some time to prevent a successful response and would have a higher probability than the product of the independent probabilities. This common cause may or may not be identified, but could for instance be a common failure in maintenance or calibration affecting all three safety devices, or a common mode of failure to which all three safety devices are susceptible, such as a loss of electrical supply. Such a loss of supply might have a probability of 1.0E−04 for instance, and clearly if it failed all three safety systems, then this is the best achievable probability of failure of the system. To represent this, the common cause failure probability would be combined with the independent probabilities using the same fault tree shown in Figure 5.5. The common mode failure, as with the HELV, would dominate the overall probability, and would suffice to correct an over-optimistic claim for redundancy.

Similar claims for redundancy in terms of human actions are often made, but in reality these can be defeated in many ways by various common causes, although the causes may not always be obvious at the time. One of the most usual ways in which supposedly independent human actions fail, is through human dependent failure, an important subject which is discussed in the following section. An example of this is the maintenance of redundant safety devices at the same time and in the same place by the same person.

5.4 Error dependency

5.4.1 Introduction

The concept of error dependency is crucial to the understanding of human error. It represents an opposing influence to error recovery and it is essential to take dependency into account when modelling human error. Error dependency is most important in the case of sequential activities such as those described in the task analysis of the batch reactor in Chapter 4. A human dependent failure is defined as *an error whose probability is increased as a result of an earlier error which made it more likely*. How a previous error can make a subsequent error more likely is discussed below.

An example of human dependent failure was mentioned in Section 3.2.4 and a case study for this is described in Chapter 9 under the form of a maintenance error.

The apparently highly reliable multiple redundancy of a four-engine BAe146 aircraft of the Royal Flight was degraded due to a series of human dependent failures. The failure was an error of omission in inserting an 'O' ring in the oil filter of each of the engines. Although the error was described in Section 3.2.4 under the heading of latent error, because it was not revealed until the aircraft had left the ground, it is also an excellent example of a dependent error. Although the probability that the technician would fail to insert the 'O' ring in the first engine might have been very small, the probability that he would go on to make the same mistake on the second engine, was higher. Once the failure had occurred on the second engine, the probability of it occurring on the third and fourth engines was even greater assuming that no human error recovery mechanism came into play. The first error made the second and subsequent errors more likely. The first error was 'the initiating error', the following errors were 'dependent errors'.

Two factors can lead to the occurrence of dependent errors. These are root causes and coupling mechanisms (Hollywell, 1994). The root cause relates to deficiencies in human performance which can cause the initiating error or the dependent errors which follow from it. Coupling mechanisms link the effects of the root cause to multiple human actions. Each is discussed in more detail below.

5.4.2 Root causes

The root causes of human dependent failure are divided into two types:

1. *Externally induced*. These are external mechanisms and performance shaping factors which have a common influence on two or more tasks leading to dependent errors which may thus be coupled together. They are broadly analogous to common cause failures in equipment items. Common cause failures are external mechanisms which have a common effect on two or more items of equipment, which may or may not be identical. While the independent probability of a failure of two or more items at the same time may be extremely small, the probability of the common cause is often much higher. The result is a failure of all the systems at the same time. The probability of the common cause failure dominates the probability of a complete system failure in the same way as shown in the fault tree for HELV in Figure 5.5. The HELV input to the fault tree is effectively the same as a common cause failure. Thus, when the probability of two or more independent errors occurring at the same time is very small, linking of the two errors by an external common cause results in an overall probability of joint failure which is much greater. This will be demonstrated in the dependency model given below.

    An example of an external mechanism which is the root cause of two dependent errors might occur if the two actions are carried out under common adverse circumstances such as:
    - poor design of human–machine interface;
    - extreme conditions in the work environment such as high temperature, humidity or noise level;
    - deficiencies in organization of the task;

– excessive task demands;
– inadequate training.

The above list is by no means comprehensive and there are many more examples of external factors which can degrade the performance of two or more tasks leading to dependent errors. External factors can influence the performance of a series of tasks carried out by the same or different individuals, since the common factor will be present at all times. This is related to the phenotype of an error or its physical outward manifestation (see Section 2.2).

2. *Internally induced.* These errors are analogous to common mode failures in equipment items although it is necessary to be cautious in stretching this analogy too far. A common mode failure is defined as *a fault condition that is specific to a particular design of equipment, so that if two identical equipment items are installed, it is possible that they will both fail at the same time from the same cause.* The best method of preventing common mode failures is by introducing diversity of design, usually by using different items of equipment to carry out the same function. In the case of human dependent failure, internally induced errors are found in the same individual carrying out similar tasks which are close together in time or place. It is sometimes called 'within-person dependency'. Strictly speaking, these types of dependent error will be caused by internal psychological error inducing mechanisms which are the experience of a particular individual at a particular time.

A typical example of an internally induced dependent error is one which occurs in a series of similar tasks where a simple slip occurs in the first task, and remains undetected by the person making the error. The probability of making an error in the second and subsequent tasks is thus increased solely because an error has occurred in the first task, but not due to any discernible external cause. The analogy with equipment common mode failures breaks down very quickly when it is attempted to discover the psychological mechanisms at work in a particular instance. In most cases the actual mechanism is indefinable. This is related to the genotype of the error or its cause at the cognitive or psychological level (see Section 2.2). Fortunately it is not necessary to be fully aware of the psychological mechanisms at work when undertaking human reliability assessments.

5.4.3 Coupling between errors

In order for dependency to occur between two errors it is usually necessary for some form of 'coupling' to exist between the two actions where the errors occurred. In the case of the incident involving the BAe146 aircraft of the Royal Flight the coupling was the fact that the tasks were identical, and that an error made the first time would be carried over into subsequent tasks. However, neither the errors nor the task have to be identical, it is only necessary that some form of coupling exists between the two. It is also important to note that in a dependent error situation, the earlier error does not necessarily make the dependent error inevitable. In most cases it will only increase the probability of the second and subsequent dependent errors compared with the probability of error if they had occurred independently.

5.4.3.1 *Coupling mechanisms*

The following is a summary of the common types of coupling mechanisms:

- Nature of task – the coupling between the two errors may be a similarity in the nature of the task. If an error occurs in the first task of a series of similar tasks, then clearly as in the example above, the same error in the second and subsequent tasks will become more likely.
- Nature of error – the tasks do not have to be similar for error dependency to occur. Coupling can occur between dissimilar sequential actions such that an error in the second action becomes more likely if the same error in the first action has already occurred. This is often due to some common influence on task performance which affects both actions in a similar way. The common influence could, for instance, take the form of an external or environmental factor which causes the person undertaking the actions to make the same mistake.
- Temporal effects – it is generally accepted that the strength of the coupling between the two dependent errors diminishes with time. Thus if two similar actions are carried out in rapid succession, an error in the first action will make the same error in the following action much more likely than if the two actions were separated in time. A time delay may temporarily break the concentration of the person carrying out the task. When attention to the task in hand is restored either the initiating error is discovered making a dependent error less likely or the correct method of carrying out the task is resumed so that the dependent error does not occur. The time delay may be caused by a distraction such as being called to answer the telephone.
- Spatial effects – if the first task is carried out in one place, but there is then a change of location before the second task is carried out, there will be a diminution in the coupling effect. This could of course also be time-related, but in general the change of location provides an opportunity, even momentarily, for the attention to the task to be refreshed. As above, this may lead to the discovery of the initiating error as well as the dependent error.

These coupling mechanisms can be used with the THERP dependency method described below to determine the level or degree of dependency between two tasks so that the increased probability of the dependent error can be estimated.

5.4.4 The importance of error dependency

Error dependency has been the cause of many serious accidents. It is extremely relevant to maintenance activities where the same person often carries out the same task on different components. In many safety critical systems, such as the trip system for a chemical reactor, the necessary high reliability for the system is obtained by providing redundant components. For example, if a chemical reactor needed to be shut down quickly due to a high temperature being reached, then the temperature sensors used to measure the temperature would be duplicated or even triplicated. Thus, if a single sensor failed to detect a rise in temperature, the other detector(s) would register the increase and cause the reactor to shut down. Not only would redundant temperature sensors be provided, but the other components needed to activate a trip would also be duplicated or triplicated. In such a case, it would be very important for these redundant components to be maintained to a high standard. The maintenance activity must therefore be as free as possible of human error.

If a human error occurred during the maintenance of one out of a set of three redundant components, and this error were not detected, it might be assumed that the provision of the other two components would still enable the trip system to operate correctly in the event of a high temperature. However, this would only be the case if the same error were not made during the maintenance of the other components. If the three components were maintained in succession by the same technician, and an error was made during the work on the first component, then the probability that the same error would be made on the other two components is increased due to the effects of error dependency described above.

In this situation, the most likely cause of failure of a sequence of activities would be an internally induced dependent failure, arising from a common psychological factor influencing the performance of the technician. The error could indeed be random in nature, as described in Section 1.3.5, such as a simple omission to install an O-ring gasket for no apparent reason except forgetfulness. Alternatively, as in the case of the BAe146 aircraft of the Royal Flight described in Chapter 9 it may be a system induced error arising perhaps from a lack of a quality control system or a failure to put this control system into action. In reality, what at first sight appear to be random errors may turn out to be system induced or dependent errors on closer examination. It may be important to understand the reason for the first error but it is equally if not more important to recognize that the commission of the first error will increase the probability of the same error occurring in the second and subsequent tasks, resulting in a multiple task failure.

5.4.5 Dependency modelling

Accounting for human dependent failures is an extremely important aspect of HRA. This is particularly the case with sequential activities, such as maintenance procedures, which are safety critical. The assumption of independence between tasks within a procedure may result in unjustified levels of confidence in the outcome. If human dependency effects are ignored, then the analysis of safety critical procedures will prove to be over-optimistic and the estimation of the level of risk, if this is the purpose of the analysis, will be understated.

The only realistic method that is currently available for modelling human dependent failure is the THERP dependency model (Swain and Guttmann, 1983). The THERP dependency model considers five levels of dependency: zero, low, medium, high and complete. For each dependency level, two types of coupling mechanism have been defined, spatial and temporal. Each level of dependency is described below.

5.4.5.1 *Zero dependency*
Zero dependency means that two potential errors within a task are completely independent and therefore the multiplication rule of probability can be confidently applied

to obtain the overall reliability. There would be no need, for instance, to apply a HELV. Zero dependency is defined by the following coupling mechanisms.

5.4.5.1.1 Coupling mechanisms

1. *Spatial (or relative distance between actions).* The actions take place in a different area. This might mean that the person carrying out the tasks has, in between the tasks, to walk to a different room or set of controls, collect required tools or parts, etc. from a different area of the building.
2. *Temporal (or time between actions).* The time taken is greater than 60 seconds. This may or may not be associated with the spatial coupling (i.e. if the time between the actions was greater than 60 seconds then even if the actions took place in the same area, the dependency level would be zero.

5.4.5.1.2 Calculation of dependency

The formulae for calculation of dependency is:

$$P_{(b/a)}/\text{ZD} = P_b$$

The formulae expresses the conditional probability $P_{(b/a)}$ of an error in Task B given that an error has occurred in Task A. ZD signifies that there is zero dependency existing. Thus Task B is effectively independent of Task A and its independent probability will remain unaffected.

5.4.5.2 *Low dependency*

5.4.5.2.1 Coupling mechanisms

1. *Spatial (or relative distance between actions).* The actions take place in the same general area. This might mean that the person carrying out the tasks has, in between the tasks, to break away to refer to instructions, retrieve different hand tools, etc., these tools being a short distance away.
2. *Temporal (or time between actions).* The time taken is between 30 and 60 seconds. Again, this may or may not be associated with the spatial coupling.

5.4.5.2.2 Calculation of dependency

The formulae for calculation of dependency is:

$$P_{(b/a)}/\text{LD} = (1 + 19P_b)/20$$

In this case the value of the conditional probability $P_{(b/a)}$ will be slightly more than the independent probability P_b.

5.4.5.3 *Medium dependency*

5.4.5.3.1 Coupling mechanisms

1. *Spatial (or relative distance between actions).* The actions take place in the same general area but also in the same visual frame of reference. This might mean that the person carrying out the tasks must perform a cognitive (thinking) exercise in

between actions which is associated with the tasks but breaks attention. In practice this might mean reading the value on a gauge or calibration instrument but without changing the main location.
2. *Temporal (or time between actions)*. The time taken is between 1 and 30 seconds which may or may not be associated with the spatial coupling.

5.4.5.3.2 Calculation of dependency
The formulae for calculation of dependency is:

$$P_{(b/a)}/MD = (1 + 6P_b)/7$$

In this case the value of the conditional probability $P_{(b/a)}$ will be considerably more than the independent probability P_b.

5.4.5.4 *High dependency*
5.4.5.4.1 Coupling mechanisms
1. *Spatial (or relative distance between actions)*. For high dependency the actions take place in the same visual frame of reference (as for medium dependency) but the person carrying out the tasks does not have to change position at all, nor will there be any delay between the tasks.
2. *Temporal (or time between actions)*. The time taken is less than 1 second.

5.4.5.4.2 Calculation of dependency
The formulae for calculation of dependency is:

$$P_{(b/a)}/HD = (1 + P_b)/2$$

In this case the value of the conditional probability $P_{(b/a)}$ will be greater than 0.5 depending on the original value of P_b. This effectively means that the dependency of Task B on Task A is such that it will fail on at least half the occasions.

5.4.5.5 *Complete dependency*
Complete dependency means that if the first or initiating error occurs, then the probability of a subsequent dependent error will be 1.0 or certainty.

5.4.5.5.1 Coupling mechanisms
1. *Spatial (or relative distance between actions)*. For complete dependency the actions are clearly in the same visual frame of reference but comprise a virtually simultaneous operation such as would occur when using one hand or two hands together to carry out two rapidly consecutive tasks with hardly any delay between the tasks.
2. *Temporal (or time between actions)*. The actions take place virtually simultaneously or in rapid succession.

5.4.5.5.2 Calculation of dependency
The formulae for calculation of dependency is:

$$P_{(b/a)}/CD = 1.0$$

In this case, whatever the original value of the independent probability P_b the conditional probability $P_{(b/a)}$ will always be 1.0 or certainty. This means that the dependency of Task B on Task A is so complete that it will always fail.

5.4.5.6 *Dependency calculation method*

The formulae given for the zero, low, medium, high and complete levels of dependency above express the probability of an error in a Task B given that an error has occurred in a preceding Task A. The symbols ZD, LD, MD, HD and CD, respectively represent the levels of dependency. The probability of an independent error in Task B, without dependency, is given by P_b. The probability of an error in the preceding Task A is P_a. $P_{(b/a)}$ therefore represents the conditional probability of an error in Task B given that an error has occurred in Task A. Thus, for instance, $P_{(b/a)}$/MD represents the conditional probability of an error in Task B given that an error has occurred in Task A, at a medium level of dependency. While the actual values obtained by using these formulae are empirical and somewhat artificial, the main advantage is that it is possible to obtain a degree of consistency in correcting for dependency across a wide range of situations.

It is important when making these calculations to remember that the error probability being corrected for dependency is the probability of the second or dependent error. There is no need to correct the initiating error probability, which is estimated in the normal way using database or other methods. Nor is there necessarily any relationship between the Task A and Task B errors. Both the tasks and the errors may be completely different. It must be understood that it is the *occurrence* of the initiating error in Task A, not its probability, that is causing the dependency to occur in Task B. This is most easily demonstrated by the simple example of a maintenance task followed by a check upon that task to detect whether an error has occurred.

5.4.5.6.1 Example of dependency calculation

A technician is required to assemble a mechanical seal on a chemical pump, following maintenance. The purpose of the seal is to prevent a toxic liquid from inside the casing of the pump leaking through the gap where the electrical drive shaft penetrates the pump casing. The seal is made by pressurizing a mechanical assembly which grips the shaft and forces any leakage back into the pump. Once the pump has been assembled, it is primed with liquid and started up in order to check that no leakage is taking place. The check is effectively a recovery action.

It is estimated that the probability of an error during the assembly of the mechanical seal is 1 in 1000 maintenance operations, or $P_m = 0.001$. The probability of a failure to detect an assembly error is estimated to be 1 in 100 checks or $P_c = 0.01$. This includes a failure to carry out the check. The check is carried out by the same technician in the same place, although he has to remove the maintenance safety lock from the local start panel. There is a time delay between the first task and the check task exceeding 60 seconds. As the second task is carried out in the same place as the first one it is estimated, rather pessimistically, that there is a low dependency between an error in the check and a maintenance task error. The assumption in this situation is that if a maintenance error occurs, a successful check needs to be carried out if the pump is not to leak when it is put back into service. If there is a maintenance error AND a failure of the checking task, then the pump will leak toxic solution. From the multiplication rule of probability, the probability of the pump leaking when it is started after maintenance (P_f) would be calculated from the product of P_m and P_c, thus:

$$P_f = P_m \times P_c = 0.001 \times 0.01 = 0.00001 = 1.0E-05.$$

As we have seen earlier, this rule assumes that the two events are completely independent. Also it may be noted that the joint probability of failure P_f is already at the suggested lower limiting value of probability set at $1.0E-05$. It has already been decided that the two events are not independent but that a low dependency state exists between them. Using the appropriate dependency formula from Section 5.4.5.2.2, the probability of the dependent error needs to be adjusted.

$$P_c/P_m/LD = (1 + 19P_c)/20 = [1 + (19 \times 0.01)]/20 = 0.0595$$

$P_c/P_m/LD$ represents the conditional probability of a failure of the check dependent upon P_m having occurred, with a low dependency level. The result is 0.0595 which can be rounded to 0.06 or $6.0E-02$ instead of 0.01 or $1.0E-02$ for the independent probability P_c. This represents a six-fold increase in probability of the checking task due to dependency.

Using the adjusted probability for P_c of $6.0E-02$, the multiplication rule can now safely be applied having taken dependency into account. The overall probability of failure (i.e. the pump seal leaking due to a failure in maintenance and checking) P_f is therefore:

$$P_f = P_m \times P_c/P_m/LD = 0.001 \times 0.06 = 0.00006 \text{ or } 6.0E-05$$

This is clearly a more realistic estimation of the joint probability having taken dependency into account and represents an alternative approach to the lower limiting value method adopted above. It illustrates the advantage of using an independent check on the technician's work by a different person. If this was done then zero dependency could be assumed and the overall failure probability could possibly be taken as $1.0E-05$. Numerous examples of dependent failure can be found in the accident case studies in Part II of this book.

References

Hollywell, P.D. (1994). Incorporating human dependent failures in risk assessment to improve the estimation of actual risk. *Proceedings of Conference on Risk Assessment and Risk Reduction*, Aston University, March 1994.

Swain, A.D. (1987). *Accident Sequence Evaluation Programme*, Washington DC: US Nuclear Regulatory Commission.

Swain, A.D. and Guttmann, H.E. (1983). *Handbook of Human Reliability Analysis with Emphasis on Nuclear Power Plant Applications. Final Report (NUREG 1278)*, August 1983, Washington DC: US Nuclear Regulatory Commission.

6

Human error in event sequences

6.1 Introduction

This chapter shows how it is possible to model and analyse, from a human error perspective, an event sequence that can lead to an accident or other undesirable outcome. There is, however, a major difficulty here. Accidents are extremely complex and usually it is only possible after an accident to understand fully the actual sequence of events that took place. Many potential accidents have been predicted in advance and some of these predictions have come true. However, it is notoriously difficult to predict the exact sequence of events which will lead to an accident prior to the accident occurring. For this reason, it is usually with hindsight that measures are put in place to prevent the sequence of events occurring again.

As there are so many potential pathways that can lead to an accident, measures taken to prevent a particular pathway recurring will not necessarily be effective in blocking other pathways. The reason is often given that not all of the potential pathways can be foreseen. This is not necessarily the case. Firstly there are many techniques available for hazard prediction, such as hazard and operability studies (HAZOPS) and various types of risk assessment (which are beyond the scope of this book to describe). Such techniques are designed not only to predict potential hazards, but also to estimate their likelihood and consequence. They may also, if taken to a sufficient level of detail, be able to identify particular tasks and activities that, if they failed, could lead to the identified hazards. If a critical task or activity is identified it is then possible to predict, using the methods described in Chapter 4, the ways in which a human error might fail this activity and lead to an accident. However, these methods are only able to identify and possibly quantify human errors occurring in isolation. What is needed is a human error modelling method that is able to represent the tasks in an activity or procedure and take account of the interactions between them. Such a method is described in this chapter. It is based on a process of event tree analysis and is capable of identifying and modelling the numerous error pathways that could lead to a failure of an activity or procedure. It explicitly models the important aspects of error dependency and error recovery. This modelling tool not only increases the level of understanding of how human errors occur within an event sequence but also indicates the pathways which make the greatest contribution to failure.

6.2 Human reliability event trees

6.2.1 Introduction

In Chapter 5, Figure 5.3, it was shown how an event tree could be used to model a sequence of events where the probability of each separate event had already been assessed. The advantage of the event tree is that the events are shown in time order and the tree immediately reveals all possible pathways to success or failure of the sequence. It is then possible to calculate the probability of a successful (or failed) sequence by multiplying together the probabilities from each node of the tree. The example event tree shown in Chapter 5, however, treated each event as independent of the other events. As discussed in the previous chapter, ignoring dependencies between human error events can lead to over-optimistic estimates of reliability. The type of event tree construction shown in Chapter 5 could represent any type of event, whether this is a mechanical or electrical failure of a component, or a human error. However, this chapter provides an alternative way of drawing event trees to represent human error in a sequence of activities. These are called human reliability analysis (HRA) event trees. It should be noted that whether an event tree is drawn in the standard way, as shown in Chapter 5, or as an HRA event tree, the method of calculation and the result in terms of overall probability of failure of the sequence is exactly the same. The advantage of the HRA event tree is a more explicit representation of human error dependency and recovery.

6.2.2 Example human reliability analysis event tree

The use of an example is the most effective way to describe the construction of an HRA event tree. Figure 6.1 (reproduced by kind permission of Electrowatt-Ekono (UK) Ltd) represents a completed tree for the task 'operator starts circulating water pump' used as an example for task analysis and error identification in Chapter 4.

6.2.2.1 *Labelling convention*

The HRA event tree diagram is shown on the left. The errors identified from the task analysis, as presented in Table 4.7 of Chapter 4, are shown in the probability table on the right. Each error is given a reference number (in the left-hand column from 1 to 8) and the outcome of each error is designated by a letter (in the next column from A/a to F/f). Lowercase letters are used to designate a successful outcome (i.e. the error does not take place and the task is completed successfully) and uppercase letters are used to designate a failure (i.e. the error has occurred). A description of the error is given in the right-hand column of the table. The probabilities of success and failure are shown in the last column corresponding to the lower and uppercase letters.

Human reliability event tree

Probability table

Event reference number	Action success / Failure	Description of error	Probability P_s / P_f
1	a	Operator proceeds to wrong switch room	1.00
	A		0.00
2	b	Operator selects wrong starter panel	0.9990
	B		1.00E-03
3	c	Operator fails to reset pump	0.9970
	C		3.00E-03
4	d	Operator fails to start pump	0.9999
	D		1.00E-04
5	e e' e"	Operator fails to check pump is running	0.9910
	E E' E"		9.00E-03
6	f	Operator checks wrong pump (with dependency)	0.9491
	F		5.10E-02
7	g g' g"	Operator fails to detect pump not running	0.9999
	G G' G"		1.00E-04
8	f' f"	Operator checks wrong pump	0.9990
	F' F"		1.00E-03

Failure probabilities (event tree):

F_1 — Negligible
F_2 — 9.00E-06
F_3 — 5.05E-05
F_4 — 9.41E-08
F_5 — 2.70E-05
F_6 — 2.97E-06
F_7 — 2.97E-07
F_8 — 8.96E-07
F_9 — 9.87E-08
F_{10} — 9.86E-09

Total P_s = 0.999909   Total P_f = 9.08E-05

Figure 6.1 HRA event tree for task: operator starts circulating water pump.

The pathways emanating from the nodes in the HRA event tree are labelled as a failure path running across (uppercase letter) or a success path running downwards (lowercase letter). The probabilities corresponding to each path are taken from the table of probabilities. Each failure path ends with an outcome probability and these are labelled F_1–F_n where 'n' is the number of failure outcomes in the tree. In this tree there are 10 possible outcomes. Thus error A, following path A, ends in outcome F_1, the next failure path represented by B–E ends in outcome F_2 and so on. The logic of the tree is described below.

6.2.3 Quantification of error probabilities

In all cases, the errors have been quantified using the technique for human error rate prediction (THERP) methodology described in Section 4.3.1.3. This method has been chosen in preference to others because THERP is particularly useful in providing error probabilities for operating tasks of the type characterizing this procedure. The errors are divided into two categories, procedural errors that take place during the commissioning of the circulating pump and recovery errors taking place during the checking of the procedure. A brief description of how each of the errors has been quantified is given below. The construction of the tree will be described in Section 6.2.4.

6.2.3.1 *Procedural errors*

6.2.3.1.1 Error *A*: Operator proceeds to wrong switch room

This error probability is considered to be negligible, since there is only one switch room, and it is therefore assigned a value of zero. The potential error is included to demonstrate completeness of the analysis.

6.2.3.1.2 Error *B*: Operator selects wrong starter panel

The design of this particular plant is such that the correct starter panel must be selected from an array of eight similar starter panels all of which are identified by clear and unambiguous labels. Referring to Section 4.3.1.3.1 in Chapter 4, the most applicable error is 'select the wrong circuit breaker in a group of circuit breakers' where more favourable performance shaping factors (PSFs) apply. The median value given by THERP is 3.0E−03, with an error factor (EF) of 3. This means that the upper and lower bounds are from 1.0E−03 to 9.0E−03. As the PSFs are considered to be highly favourable, an error probability of 1.0E−03 at the lower bound has been selected.

6.2.3.1.3 Error *C*: Operator fails to reset pump

This is considered to be an error of omission from an item of instruction in a written procedure. THERP suggests a range of probabilities for this error (see Section 4.3.1.3.2 in Chapter 4). In this case a written procedure is used with a list of instructions consisting of less than ten items. A check is later carried out to ascertain that the pump is running and this represents a potential recovery factor. As an event tree is being used, it is possible to show the check explicitly and it is represented as a separate task (see events *E*/*e*, *F*/*f* and *G*/*g* below). In Section 4.3.1.3.2 the median probability of error is 3.0E−03. The median value is used because the PSFs are considered to be average.

6.2.3.1.4 Error *D*: Operator fails to start pump

This error of commission (ECOM) is one which is immediately obvious once it has been made. The starter handle is spring loaded and when it is depressed there is an immediate feedback due to the vibration of the starter solenoids engaging and a reading appearing on the ammeter. There is no particular reference in the THERP methodology to this type of error but because is so easy to detect, a low probability of 1.0E−04 is assigned. An even lower probability of error might have been assigned to this task but was avoided because there is a further possibility of recovery when the check that the circulation pump is running is carried out later.

6.2.3.2 *Recovery errors*

As shown in the event tree, the recovery actions *E*/*e*, *F*/*f* and *G*/*g* provide the potential for recovery of all three procedural errors *B*, *C* and *D*. However, it is possible for the probabilities of failure of these checks to differ depending upon which procedural error is recovered. This is mainly to allow for the effects of any potential dependency of a recovery error upon a procedural error. As discussed in Chapter 5, the effect of such a dependency would be to increase the probability of the recovery error over and above its basic probability if it were independent. For this reason, different symbols are used in the tree to identify the recovery actions for each of the procedural errors, namely E/e, F/f and G/g for error B; E'/e', F'/f' and G'/g' for error C and E''/e'', F''/f'' and G''/g'' for error D. They are quantified as follows.

6.2.3.2.1 Error E: Operator fails to check circulation pump is running

Since the checks are carried out by following a written list, and because there is no further check carried out, 'an error of omission when written procedures are used' is applicable. The median value of probability for this is given as 3.0E−03 with an EF of 3 (see Section 4.3.1.3.2). Due to the time restriction upon these activities, there is an incentive to omit this check and return to the control room to continue the shutdown process. On the other hand, the ammeter on the starter panel provides a very simple indication that the circulation pump is running and would be difficult to miss. Nevertheless, a pessimistic view is taken and the upper bound probability of 9.0E−03 is used in this case. No dependency between error E and errors B, C or D has been identified.

6.2.3.2.2 Error F: Operator checks wrong pump

This error assumes that error 'E' has not occurred and the operator has in fact checked that the circulation pump is running (i.e. the checking task is successful represented by 'e' as opposed to error 'E'). However in this case the operator has selected the wrong pump to check. The error is therefore considered to be a selection error for an item of equipment which is selected from among similar items as described in Section 4.3.1.3.1 in Chapter 4. It is assumed that the selection error data quoted for locally operated valves is also applicable to a pump selected from a group of similar pumps. The closest match is a component which is clearly and unambiguously labelled and set apart from items that are similar in all of the following respects: size and shape, state and presence of tags. The median error probability in Section 4.3.1.3.1 is therefore 1.0E−03. The median value is used since there are no particularly favourable or unfavourable PSF's. This nominal value is adopted for recovery errors F' and F''.

However, for recovery from error B, it is considered that a low dependency (LD) between F and B exists. The probability originally assigned to the error F without any dependency is, as above, 1.0E−03. The formula for LD in Section 5.4.5.2.2 of Chapter 5 is:

$$P_{(b/a)}\text{LD} = (1 + 19P_b)/20$$

Thus the probability of F, *with* dependency is as follows:

$$F/B/\text{LD} = (1 + (19 \times 1.0\text{E}-03)/20 = 5.1\text{E}-02$$

This is the value shown in the probability table for error F. It can be seen that the probability has increased by a factor of 50 to allow for the dependency.

6.2.3.2.3 Error G: Operator fails to detect that the pump is not running

This error assumes that the operator has checked the correct pump but has failed to detect that the pump is not in fact running. This is not an error of omission since the checking task has in fact been carried out and the failure comes at the end of this check. It is possible for the operator to check by observation of the ammeter at the starter panel whether the pump is running or a more reliable check could take place at the pump itself. However since the check has proceeded this far and is nearly complete, it is highly likely that it will be completed and unlikely that the operator will not observe a reading (or lack of a reading) on the ammeter. Since there is no applicable data for this error in the data tables, and because it is unlikely, it is assigned a low probability of 1.0E−04.

6.2.3.3 *Important note on assigning error probabilities*

It may be considered that the process of assigning error probabilities above is rather imprecise and somewhat subjective. The arbitrary assignment of a low probability value of 1.0E−04 for errors D and G are examples of this. As discussed in Chapter 4, the matching of error data to actual situations is always fraught with difficulty. It is almost impossible to be sure that the performance shaping factors which influenced the error data at the point of collection have all been recorded and are fully understood. In the same way the factors which influence performance in the actual error situation may differ considerably from those existing at the point of collection of the data. However, as long as this difficulty is understood, then for situations where error matching is difficult or impossible, this fact must be recorded and any assumptions which are made in assigning an error probability should be stated.

It will be shown below how the impact of individual human errors upon the overall probability of failure of the procedure can be assessed using scenario analysis. Where individual human errors are shown to have a high impact on the overall failure probability, then these can be revisited in order to refine the original probability estimates. Individual human errors which clearly have a negligible impact on the overall failure probability can be ignored.

6.2.4 Event tree logic

6.2.4.1 *Event A/a*

The sequence of errors in the HRA event tree commences at the top left-hand corner with the first potential error No. 1 designated A/a. In the case of this error, the probability that the operator will proceed to the wrong switch room is considered to be negligible, and therefore 'A' is assigned a value of zero. Hence, if this error does not take place, then the path follows 'a' and passes down to the next node B/b which represents either a failure represented by error B occurring, or else success, error B not occurring, represented by 'b', in which case the path continues downwards to the node C/c.

6.2.4.2 *Event B/b*

6.2.4.2.1 Description of logic

Continuing along path B, which assumes that the error 'operator has selected the wrong starter panel' has occurred, then the next node to be encountered, is node E/e. Node E/e represents the possibility of recovery taking place. If the error 'operator has selected the wrong starter panel' occurs then the error will be discovered when the operator checks whether the pump is running later in the sequence. The error 'operator fails to check the pump is running' is designated as 'E'. This is therefore equivalent to an AND gate, so that for the error 'operator has selected the wrong starter panel' to fail the complete sequence, the error 'E', 'operator fails to check the pump is running', must also occur. The product of 'E' and 'B' gives the joint probability using the multiplication rule. However, it is also necessary that 'a' has occurred to reach F_2 along this path. Hence, if both 'E' and 'B' occur, given that 'a' has occurred, then the outcome is F_2, calculated from:

$$F_2 = a \times B \times E = 1.0 \times 1.0E{-}03 \times 9.0E{-}03 = 9.0E{-}06$$

We have now followed the failure path, 'a–B–E' to reach F_2. However, at node E/e it is possible that the operator checks that the pump is running, but checks the wrong pump (and finds that it is running). In this case path 'e' is followed from this node, leading to node F/f (actually shown as $(F/B/LD)$ which is explained below). F/f represents the possibility that the operator, having checked that a pump is running (e), has in fact selected the wrong pump to check. This was originally assigned a probability of $1.0E{-}03$ in the table of probabilities (as in errors F' and F'', see below). The outcome from this path is represented by probability F_3. However, it has also been assessed that a dependency exists between error F and error B and this will increase the probability of F to greater than $1.0E{-}03$. This dependency is due to a similarity in the two errors, such that if the operator selects the wrong starter panel, and goes ahead to start the wrong pump, this will increase the probability that when he finally comes to check that the pump is running, it will be the wrong pump which is checked. That is, since he believes he has started the pump, he will more naturally search for a pump that is running that one that is stopped, with an increased probability that he will identify the wrong pump.

Thus error F, operator checks wrong pump, is made more likely because error B has occurred, 'operator selected the wrong starter panel'. However, the coupling between the two errors is considered to be weak, and only a LD has been assigned. The dependency situation is represented by assigning the symbol $F/B/LD$ for the error F, meaning F is dependent upon B with a LD level (see dependency calculation for the quantification of this error under error F above).

Thus, the failure path leads via 'a–B–e–$F/B/LD$' to F_3, the overall probability of the outcome of this path. As before, F_3 is calculated from the product of the probabilities along this path thus:

$$F_3 = a \times B \times e \times F/B/LD = 1.0 \times 1.0E{-}03 \times 0.991 \times 5.1E{-}02 = 5.05E{-}05$$

Given that 'e' and 'f' occur, that is, the operator checks that the pump is running and it is the correct pump, there is still the possibility that he fails to detect that it is not running. This would lead to a failure of the overall sequence, and to outcome F_4. The probability F_4 is calculated in the same way. The failure path is now 'a–B–e–f–G' to F_4 so the probability F_4 is calculated from:

$$F_4 = a \times B \times e \times f \times G = 1.0 \times 1.0E{-}03 \times 0.991 \times 0.9491 \times 1.0E{-}04$$
$$= 9.41E{-}08$$

However, if the operator is successful in detecting the pump is running, represented by 'g', then the path follows the sequence 'a–B–e–f–g' and this is a success path that leads back to the main success path on the left of the tree. However, it is more accurate to call this return route to a success path, a 'recovery path'. Recovery paths are represented by broken horizontal lines as shown.

6.2.4.2.2 Result for event *B/b*

The overall result from the event tree so far is that the error B, the selection of the wrong starter panel, can be recovered by any of the three check tasks, represented by nodes E/e, F/f and G/g. However, a failure in any one of these tasks can lead to error B failing the whole sequence via the outcomes F_2, F_3 or F_4. Thus success is required in all three check tasks for recovery from error B to occur. It should be noted that it is not a necessary condition for these three checks to occur immediately after the task of selecting the starter panel. The checks may be effective even if they are delayed and other tasks intervene. In this sense, the event tree is not an exact representation of the order in which tasks take place. The check tasks in fact do occur after the final operator task of actually starting the circulation pump represented by the event D/d. However the same three check tasks happen to be a check upon all three of the main tasks in the sequence, events B/b, C/c and D/d. It is necessary however to calculate the effect of these tasks separately upon each of these events. The same three check tasks therefore appear three times in the tree, and are distinguished by means of prime marks, $'$ and $''$, so that event E/e appears again as E'/e' and E''/e'', as do events F/f and G/g. This will become clear when events C/c and D/d are discussed below.

It is also important to mention here that the time available to complete the tasks has an influence upon the logic of the tree. It has already been stated in the description of this activity in Chapter 4 that there is a time limit within which the circulating pump must be started if the release of toxic gas into the environment is to be avoided. Some of the errors shown in the tree such as the error B 'selection of the wrong starter panel' could of course be recovered given sufficient time for the error to be detected. The same applies to the check task errors E, F and G. Of course, the overall outcome of interest is that a failure of the operating procedure allows the toxic gas release. Thus it is not just the failure of the operating procedure that is of interest but the failure within a time scale. This is why it is necessary for HRA to be addressed in a much wider context than that of the tasks and activity alone. The other events in the tree are now described below.

6.2.4.3 *Event C/c*

Event C/c represents the operator task of resetting the pump, given that the correct starter panel has been selected (event B/b). The recovery path followed so far is 'a–B–e–f–g', leading via the broken line to event C/c. However, logically, this is exactly the same as the path as 'a–b' leading to event C/c, and for simplicity this path is used to calculate the outcomes from event C/c. Thus the broken line represents the recovery path back to the main success path, and is shown broken to signify that this is not a calculation route. This is a general convention used for drawing and calculating HRA event trees.

As stated above, the same three check tasks as used for event B/b, also apply to events C/c and D/d. Although they are actually carried out later in the sequence, after task D/d, they appear in the tree at the point in the sequence where they are applicable. Thus, if the operator fails to reset the pump, represented by error C, then the operator checking whether the pump is running, event E/e, later in the sequence, is a possible error recovery mechanism from error C. If the operator fails in this check, then error E' occurs and the failure path $a–b–C–E'$ leads to failure probability F_5. As before, F_5 is calculated from:

$$F_5 = a \times b \times C \times E' = 1.0 \times 0.999 \times 3.0E{-}03 \times 9E{-}03 = 2.7E{-}05$$

In exactly the same way as described for event B/b above, although event E/e may have been carried out correctly, and it is checked that the pump is running, it is possible that the wrong pump has been selected. This is represented by event F'/f' and if error F' occurs, then the failure path leads via $a–b–C–e'–F'$ to the outcome failure probability F_6. In this case, it is considered that there is no dependency between event F'/f' and event C/c (as there was previously between event F/f and event B/b), because of a lack of similarity between the two tasks. Error F' then takes its basic probability (without dependency) of $1.0E{-}03$. The probability F_6 is calculated in exactly the same way, using the path $a–b–C–e'–F'$, to give:

$$F_6 = a \times b \times C \times e' \times F' = 1.0 \times 0.999 \times 3.0E{-}03 \times 0.991 \times 1.0E{-}03$$
$$= 2.97E{-}06$$

Again, much as described above, the check task represented by event G'/g' 'operator failing to detect the pump is not running' is a recovery mechanism for event C/c. The pathway is $a–b–C–e'–f'–G'$, leading to outcome F_7, calculated by:

$$F_7 = a \times b \times C \times e' \times f' \times G'$$
$$= 1.0 \times 0.999 \times 3.0E{-}03 \times 0.991 \times 0.999 \times 1.0E{-}04 = 2.97E{-}07$$

If the operator does detect that the pump is not running, represented by g'', then error C has been recovered and the recovery path, shown as a broken line, back to the main success path is followed.

6.2.4.4 *Event D/d*

Event D/d represents the operator task of starting the pump. Given that the operator has found the correct switch room, selected the correct starter panel and reset the correct pump, then there is a possibility that the operator fails to actually start the pump. This may for instance be due to a sudden distraction. The path to this error on the event tree is represented by $a–b–c–D$. As with events B/b and C/c, event E''/e'', the operator checking whether the pump is running, is the first recovery action. If this check fails, then it will not be discovered that there has been a failure to start the pump and outcome F_8 results. The probability F_8 is calculated in exactly same way as those for earlier outcomes, by multiplying the probabilities along the path to failure, in this case:

$$F_8 = a \times b \times c \times D \times E''$$
$$= 1.0 \times 0.999 \times 0.9970 \times 1.0E{-}04 \times 9.0E{-}03 = 8.96E{-}07$$

Even if the operator checks that the pump is running, via success path e'', then it is still possible that he actually checks the wrong pump this error being represented by F'' leading to outcome F_9. If the operator checks the correct pump, then again, it is possible that he fails to detect the pump is not running which is represented by error G''.

The probabilities F_9 and F_{10}, are then calculated in the same way as before:

$$F_9 = a \times b \times c \times D \times e'' \times F'' = 1.0 \times 0.999 \times 0.9970 \times 1.0E{-}04$$
$$\times 0.991 \times 1.0E{-}03 = 9.87E{-}08$$

$$F_{10} = a \times b \times c \times D \times e'' \times f'' \times G'' = 1.0 \times 0.999 \times 0.9970 \times 1.0E{-}04$$
$$\times 0.991 \times 0.999 \times 1.0E{-}04$$
$$= 9.86E{-}09$$

It will be noted that although the recovery actions E/f, F/f and G/g are the recovery actions for error B, in the event tree they are shown as separate errors using the identifiers $'$ and $''$ for errors B and C. This is to allow for the possibility that although the actions may be the same, the probability of an error may be different due to dependency effects. It will be seen that the only case where this applies in this example is for error F.

6.2.4.5 Overall probability of failure of the procedure

Any one of the outcomes F_1–F_{10} can lead to a failure of the procedure for starting the circulating water pump. In logical terms this means that any of F_1 OR F_2 OR F_3 and so on to F_{10} lead to failure. Simply expressed, outcomes F_1–F_{10} pass through an OR gate leading to P_f the overall probability of failure which is therefore calculated by using the addition rule. From Chapter 3 it was shown that:

$$P_{(a \text{ or } b)} = (P_a + P_b) - (P_a \times P_b) = (P_a + P_b) - P_{ab}$$

and similarly,

$$P_{(a \text{ or } b \text{ or } c)} = (P_a + P_b + P_c) - (P_a \times P_b \times P_c) = (P_a + P_b + P_c) - P_{abc}$$

However, it can be seen that when the values of the probabilities are very small, or when the number of events passing through the OR gate are very large, the product term (P_{abc}) becomes extremely small compared with the sum, and can be ignored. In this case there are ten separate probabilities, and therefore to all intents and purposes the overall probability of failure (P_f) of the procedure is equal to the sum of the separate probabilities. In this case the sum (P_f) of F_1–F_{10} is seen to be 9.08E$-$05. In approximate terms this is a probability of failure of 1.0E$-$04 or about 1 in 10,000 operations. The success probability is of course $1.0 - 9.08E{-}05 = 0.999909$.

6.2.4.6 Note on calculation method

The most practicable way to develop and solve a HRA event tree is by using a spreadsheet package such as Microsoft EXCEL© which was used to produce Figure 6.1. This not only achieves more accuracy in solving the mathematics of the tree, but also enables rapid 'what if' calculations to be made enabling the effects of changes to the procedure to be analysed. This is described in more detail in Section 6.3, Scenario analysis.

6.2.4.7 Note on use of human error limiting values

It will be noted that some of the probabilities of the outcomes F_1–F_{10} are extremely small, and certainly many of them are much less than the human error probability (HEP) limiting value of 1.0E$-$05 suggested in Chapter 5. All the basic HEP values for the procedural tasks are greater than 1.0E$-$04 (1 in 10,000) and can therefore be considered reasonably achievable. However, because of the recovery actions and subsequent multiplication of individual task probabilities by the recovery factors, the resulting outcomes are often very small. While it would be possible to apply the limiting value to these outcomes, greater utility is obtained from the event tree by examining the relative contribution of each of the outcomes to the overall probability. The application of a limiting value would in this case to a large degree obscure the information which can be gained from the tree by exploring the outcomes of the various scenarios. The method of undertaking this analysis is described in Section 6.3.

6.2.4.8 Contribution of failure paths to overall probability of failure

The relative contributions of each of the separate outcomes to the probability of failure of the overall procedure have been calculated and are shown in Table 6.1.

From Table 6.1 it can be seen immediately that the highest contributor to the overall failure probability is the outcome F_3 which contributes over half of the total failure probability, followed by the outcome F_5 which contributes nearly a third. Outcome F_3 represents the scenario whereby the operator selects the wrong starter panel, proceeds to check that a pump is running, but then checks the wrong pump. There is a hidden assumption to be noted here, that the pump that the operator checks is actually running, although it is not the pump he should be checking. This provides a deeper insight into the logic of the tree and it may need to be explored in more detail whether the scenario is in fact realistic. For instance, it is possible that the operator will wrongly select a panel where a pump is running which could be misidentified as the circulation pump. This could be the case for instance, if there was an identical or similar looking panel nearby which could feasibly be selected by mistake.

The next highest contributor, outcome F_5, represents the scenario that the operator has failed to reset the circulation pump and a check that the pump is running, has failed. There is no assessed dependency in this situation, and the probability of F_5 at 2.7E−05 is a major contributor because it is the product of two relatively high probabilities, that of error $C = 3.0E−03$ and especially the recovery factor, error $E' = 9.00E−03$, which is approaching 1.0E−02 or 1 failure every 100 checks.

Table 6.1 Contribution of human error outcomes to overall failure probability

Outcome	Probability	Contribution (%)
F_1	0.00E+00	0.00
F_2	9.00E−06	9.91
F_3	5.05E−05	55.58
F_4	9.41E−08	0.10
F_5	2.70E−05	29.70
F_6	2.97E−06	3.27
F_7	2.97E−07	0.33
F_8	8.96E−07	0.99
F_9	9.87E−08	0.11
F_{10}	9.86E−09	0.01
Total	**9.08E−05**	**100.00**

6.3 Scenario analysis

6.3.1 Introduction

The most systematic approach to identifying weaknesses in the procedure is to undertake a 'scenario analysis' by setting up a number of hypothetical cases and carrying out 'what if' calculations. In this way it is possible to identify why the major contributors are making a high contribution to the overall failure rate. Using that information, it may be possible to make changes to the procedure to reduce the probabilities of the major contributors to failure and thus reduce the overall failure probability of the procedure. In order to demonstrate this, two scenario analyses have been carried out in order to investigate the influence of the main contributors to failure, outcomes F_3 and F_5.

6.3.2 Scenario 1: Influence of dependency on outcome F_3

If the outcome F_3 is indeed realistic, then it needs to be assessed whether the high contribution of this outcome to the overall failure rate can be reduced by changes to the operating procedure. In this case, it is fairly obvious that the high contribution has been caused to a large degree by the existence of a dependency between errors F and B, albeit a LD. It was stated above that this dependency was due to a similarity in the errors, that is if the operator selected the wrong panel and checked for the circulation pump running, there is an increased probability that he would check the wrong pump (and find it was running). If this dependency did not exist, then it can quickly be calculated that the outcome F_3 would have a probability of only $9.91E-07$. This is calculated in exactly the same way as described above but substituting the basic value F' or $F'' = 1.0E-03$ (without dependency) for the value of $F = 5.10E-02$ (with dependency). This is combined with other outcomes as shown in Table 6.2. The result is that F_3 now becomes $9.91E-07$.

It is seen that outcome F_3 now only contributes to 2.4 per cent of the overall failure probability and as might be expected the overall probability is now reduced to $4.13E-05$, about half of its previous value. Methods of reducing the dependency may have to be investigated in order to achieve this reduction. For instance, if it is indeed likely that the operator could mistake the starter panel, then some unambiguous feedback should be provided to ensure that if any check that the circulation pump is running is made upon the wrong pump it becomes immediately obvious. In this way the operator's attention is directed to his previous error of selecting the wrong panel.

6.3.3 Scenario 2: Influence of recovery factor in scenario F_5

As shown in Table 6.1, the outcome F_5 contributes about 30 per cent of the total failure rate as calculated from the event tree. It is worth investigating whether it is possible to reduce this contribution by improving the procedure. The main reason why F_5 is a high contributor, is the recovery factor E' which has the relatively high failure probability

Table 6.2 Scenario 1: Zero dependency in outcome F_3

Outcome	Probability	Contribution (%)
F_1	0.00E+00	0.00
F_2	9.00E−06	21.78
F_3	9.91E−07	2.40
F_4	9.41E−08	0.23
F_5	2.70E−05	65.25
F_6	2.97E−06	7.19
F_7	2.97E−07	0.72
F_8	8.96E−07	2.17
F_9	9.87E−08	0.24
F_{10}	9.86E−09	0.02
Total	**4.13E−05**	**100.00**

Table 6.3 Scenario 2: Improved recovery factor for E, E' and E''

Outcome	Probability	Contribution (%)
F_1	0.00E+00	0.00
F_2	3.00E−06	17.88
F_3	9.97E−07	5.94
F_4	9.46E−08	0.56
F_5	8.99E−06	53.60
F_6	2.99E−06	17.81
F_7	2.99E−07	1.78
F_8	2.99E−07	1.78
F_9	9.93E−08	0.59
F_{10}	9.92E−09	0.06
Total	**1.68E−05**	**100.00**

of 9.00E−03 (nearly 1 in 100). As discussed above, the median value for this recovery error is 3.0E−03 with an EF of 3, but because of the time restriction it is believed that there is an incentive to omit this check. The median value is therefore multiplied by 3 to give the assessed probability of 9.0E−03. In order to reduce the contribution of this error to the overall probability, it might be useful to examine the reasons for the time restriction. In fact, it would quickly be found that the time restriction is due to the process conditions and is not easily corrected, except possibly by some redesign of the process. To see if this is worthwhile, it is necessary to run the scenario analysis. If the time restriction could be corrected, then it might be possible to adopt the median value of this error and the overall failure probability would be reduced. The reduction can be calculated in the same way as above using the 'what if' scenario of replacing the assessed value with the median value in the event tree. As the time restriction applies to the recovery from any of the errors A, B or C, then the reduced probability applies to all three of the recovery errors E, E' and E''. This produces an improvement to all of the outcomes F_2, F_5 and F_8. It is assumed that Scenario 1 is also in place, using zero dependency for outcome F_3. The result is shown in Table 6.3.

This result is a further reduction in the overall failure probability from 4.13E−05 to 1.68E−05, a reduction to about 40 per cent of the previous value. It is of interest that the outcome F_5 is still the highest contributor and the reduction in probability is quite small. Similarly the reduction in probability of outcomes F_2 and F_8 is also small, although they were never major contributors. It is probable that a redesign of the process to remove the time restriction may not be cost effective in terms of the improvement to the procedure.

If both of the improvements identified in Scenarios 1 and 2 could be implemented, then the overall failure probability would be reduced to 18 per cent or about 1/5th of the original probability. Scenario analysis is a powerful tool in identifying the major human error contributors to overall failure and exploring ways in which the overall failure probability might potentially be reduced. In some cases, as for outcome F_5 above, it will be found extremely difficult to make a significant reduction in failure probability. However, without undertaking an event tree analysis of this sort, it will be extremely difficult to even identify the existence, never mind the magnitude, of the contribution of the main human errors causing failure.

6.4 Overview of human error modelling

6.4.1 Introduction

One of the main purposes of representing a procedure in the form of an HRA event tree is to identify vulnerabilities where human errors are more likely to occur. Using scenario analysis, it is possible to investigate how these vulnerabilities may be corrected so that the reliability of the overall procedure can be improved. It enables human errors to be understood not as individual failures but in the wider context of a complete activity, taking into account the performance shaping factors as well as error recovery and dependency. Analyses such as those carried out in this chapter and developed in preceding chapters are useful where procedures critical to the safety of major hazard operations have been identified. Both maintenance and operational procedures are equally amenable to this type of analysis. This section aims to provide a summary of the stepwise process of building up and collating the information which is finally combined in an HRA event tree.

6.4.2 Task analysis

The first step in the process is to decompose activities or procedures down to an appropriate level of detail using task analysis. The level of detail to which the activity is broken down must correspond to the level of detail used in the checklist for error identification. This in turn should match the level of detail of available error probability data.

Task analysis is also used to collate and present all of the relevant information concerning an activity or procedure in a logical and consistent format. If the activity is current then the information can be based on actual practice and developed by means of observation of tasks and examination of written procedures. The latter should always be confirmed by actual observation of work practices in case the procedures are not being followed. Alternatively task analysis can be carried out during the design of plant or equipment by attempting to predict the activities necessary to operate the systems. Once the equipment is operating it may be necessary to observe the activities in order to confirm the original predictions.

Task analyses can be presented either hierarchically or sequentially. For the purposes of HRA, the sequential presentation is more useful. It also enables tasks to be set out in time order. This is extremely useful during the processes of error identification, quantification and modelling because it allows interactions between tasks which are close together in time and place to be better understood. During the preparation of the task analysis, information is also gathered about PSFs which will determine the human reliability which is realistically achievable. At this stage, it is not always necessary to understand the PSFs in any great detail. Later in the analytical process, it will become apparent which tasks are important to the success of the overall activity. It may then be necessary to revisit the task analysis and develop the PSFs in more detail. This can then be used to refine the quantification of error probability. This will introduce more certainty into the identification of particular tasks as major error or risk contributors, in order to identify improvements which might be made to the PSFs to reduce the error probability.

6.4.3 Error identification

At this stage of the analysis, it is only necessary to identify potential errors within individual tasks taken as independent events, and not necessarily in relation to other tasks or in the context of the complete activity or procedure. It is only when the procedure is modelled as a whole, much later, that the context will be considered and the full significance of recovery and dependency aspects will become clear. This illustrates a principle of economy which it is wise to adopt in HRA. In the early stages of an analysis, the identification and quantification of errors is a screening process in which the level of detail is sufficient merely to reveal which errors are critical and which are not. Later, the analysis will be carried out with more rigour but restricted to tasks and errors that are shown to be critical. HRA can be a resource intensive process when carried our rigorously so this effort must be reserved for errors which are likely to be significant.

The so-called 'generic approach' to error identification was used in the worked example in Chapter 3. This approach applies a checklist of error types, such as EOMs, ECOMs, errors of substitution, etc., to each individual task, remembering that more than one type of error may occur in a single task. This approach is considered to be the simplest and most effective way of providing confidence that all the main errors have been identified.

6.4.4 Error quantification

Approaches to error quantification described in Chapter 4 included both database and expert judgment methods. Two database methods were described, the THERP and human error assessment and reduction technique (HEART) methods. For the purposes of the worked example, quantification of the identified errors was carried out in this chapter using the THERP method. However, quantification was only attempted once the procedure had been modelled using the human reliability event tree as described above. This is so that a full understanding is gained of the interactions between the errors. For example, some tasks were identified as recovery tasks and therefore needed to be located on the recovery branches of the HRA event tree.

The THERP and HEART methods presented in Chapter 4 were developed in the 1980s and might therefore appear to be outdated. The fact is that very little useful development in the field of quantified HRA has taken place since these methods were first proposed and nothing better has become available since that time. HRA is still used in a number of high hazard industries such as the nuclear and chemical industry and where it is used it tends to draw heavily upon these two methods. More recently the HEART method has tended to be favoured. Expert judgment techniques, while having their uses, tend to be highly resource intensive and the confidence in the results is not really proportionate to the input required.

As the exercise of undertaking a HRA is time consuming and labour intensive it is often queried whether it can be justified. The need for quantification will depend largely upon the importance attached to the activity and the type of information that is being sought from the analysis. In many cases, the only requirement is to assess whether critical tasks in the sequence are accompanied by appropriate recovery actions to prevent a total failure of the event sequence as a result of a single human error. For instance, if it were found for a safety-critical procedure that an uncorrected human error with a high probability can lead to a total failure then the overall probability of failure of the procedure will be at least as great as the probability of failure of that human error. In the example above the probability of many of the procedural errors are as high as $1.0E-03$ (a typical HEP for many common tasks). If any single failure remains uncorrected the probability of failure of the procedure will be at least $1.0E-03$ (or 1 error in 1000 tasks on average) plus the probability of failure of all the other contributing errors (including their recovery factors if any). If there were two of these errors, each with a probability of $1.0E-03$, without recovery, then the failure probability of the procedure will as a minimum be $2.0E-03$ (or 1 in 500 tasks). Without a detailed quantification, but using generic values of human probability (based for instance on the ranges given for SRK based types of error) a very rapid screening estimate of the overall failure probability can quickly be obtained.

6.4.5 Human reliability event trees

The modelling of the sequence of activities using an HRA event tree, with or without quantification of error probability, is useful in identifying the various failure paths which can lead to a failure of the whole sequence. The HRA event tree also enables identification of possible recovery mechanisms and perhaps more importantly the absence of such mechanisms. It is where there is an absence of recovery that the sequence is most vulnerable to failure. In addition, because the HRA event tree is drawn in time order, it enables dependency effects to be identified. By reference to the task analysis, it is possible to check whether consecutive tasks that may be susceptible to dependency, are subject to coupling mechanisms such as the time interval between tasks, similarity between tasks and whether they are carried out in the same place.

In order to draw up the event tree logic, it is often necessary to make simplifying assumptions. As the event tree logic is refined, it may be found that some of these assumptions are in fact unjustified and need to be refined for the tree to more accurately represent reality. Once a tree has been drawn up it is quite common for the logic to reveal hidden anomalies in the procedure that have not been noticed before. While it may appear from the example used here that the overall objective of preparing an HRA event tree is to quantify the probability of failure of the procedure, the reality is often different. It is the process of analysis rather than the results which can produce the greatest benefit in terms of an increased understanding of potential factors which may undermine the reliability of a procedure.

A subject not addressed in this chapter due to lack of space, is the possibility of combining the outcomes of a number of HRA event trees (say for a number of procedures) within a standard event tree with horizontal branches (such as that shown in Figure 5.3) drawn up at a higher level. An example of this might be the maintenance of a complex item of equipment where different components are subject to a number of maintenance procedures, possibly carried out at the same time. In the example of the operating procedure used in this chapter, the activity described is only one activity in a whole sequence of activities necessary to prevent the release of toxic gas into the building. Each activity would therefore be analysed separately as shown above, and the outcomes fed into an overall tree. This is especially useful where human error and equipment failures need to be combined within the same tree. However, a description of these techniques is beyond the scope of this book.

To summarize, the main use of HRA event trees is for the representation and analysis of diverse human activities undertaken with or without procedures to identify any vulnerabilities to human errors. They can be used in the analysis of operational or maintenance activities, where there is a prescribed mode of action. HRA event trees cannot, however, be easily adapted to analyse management and decision-making activities which are the root cause of many of the major accidents described in Part II of this book. In spite of this they do have certain uses in post-accident analysis.

II

Accident case studies

The case studies of major accidents presented in Part II are chosen to illustrate the human error principles described in Part I. The accidents are taken from a wide range of industries and are grouped under chapter headings by error type or form. The error types are based on a classification described in Part I for violations, latent and active errors but also demonstrate the principles of dependency and recovery. The error forms represent the way the errors are manifested in the case study and include organizational, cultural, management, design, maintenance and operational errors. All the errors described are systemic in line with the overall objective of the book to show how most errors are triggered by underlying system faults rather than being random events or due to negligence. In many of the accidents it is shown how blame was wrongly attributed resulting in system faults remaining uncorrected. The level of detail and background information provided here is sufficient to understand both the context of the errors and the deficiencies in the systems.

Most of the case studies conclude with a human error analysis comprising a summary of the direct, root and contributory causes in terms of error types or system faults with recommendations on how the probability of the errors might have been reduced. The direct cause is usually a human error while the root cause tends to be the main underlying system fault that made the error possible or more likely. The contributory causes are less significant factors or systems that also influenced the probability of error. The system of interest may be an item of equipment, a procedure or an organizational system. For some of the case studies, a quantitative analysis is also made using techniques such as fault tree and event tree analysis as described in Part I. The objective is to provide a practical illustration of how these techniques can be used to model interactions between errors in a sequence of tasks and to identify the relative contribution of individual errors to failure of the overall activity. The table below summarizes the case studies used in each chapter and indicates the error types and forms represented.

Summary of case studies presented in Part II

Chapter	Title	Case study	Industry	Human error		Form	
				Type		Form	
7	Organizational and management errors	The Flixborough Disaster	Chemical	Latent		Management	
		The Herald of Free Enterprise	Marine	Latent		Cultural	
		Privatization of the railways (UK)	Railway	Latent		Organizational	
8	Design errors	The Grangemouth fire and explosion	Chemical	Latent		Design	
		The sinking of the ferry 'Estonia'	Marine	Latent		Design	
		The Abbeystead explosion	Water	Latent		Design	
9	Maintenance errors	Engine failure on the Royal Flight	Aviation	Latent		Maintenance	
		The Hatfield railway accident	Railway	Latent		Maintenance	
		The Potters Bar railway accident	Railway	Latent		Maintenance	
10	Active errors in railway operations	Clapham Junction accident	Railway	Latent		Maintenance	
		Purley accident	Railway	Active		Operational	
		Southall accident	Railway	Active		Operational	
		Ladbroke Grove accident	Railway	Active		Operational	
11	Active errors in aviation	Loss of flight KAL007	Aviation	Active		Operational	
		The Kegworth accident	Aviation	Active		Operational	
12	Violations	The Chernobyl accident	Nuclear	Violation		Operational	
		The A320 crash at Mulhouse	Aviation	Violation		Operational	
13	Incident response errors	Fire on Swissair flight SR111	Aviation	Active		Operational	
		The Channel Tunnel fire	Railway	Active		Operational	

7

Organizational and management errors

7.1 Introduction

This chapter provides cases studies of systemic errors arising in organizational and management systems. As discussed in Chapter 3, this type of error is nearly always of the latent type. The errors can occur at almost any level of an organization, but tend to take place within operational management positions involving executive decisions. However, these decisions may in turn be influenced by cultural factors originating at company director level. The examples given in this chapter are taken from the chemical, marine and railway industries. The chemical industry accident occurred in 1974 at Flixborough in the county of Lincolnshire. It was crucial in bringing about important changes in the way accidents are understood. It also emphasized the need for non-prescriptive, less regulation based safety legislation able to identify a much broader range of potential hazards. The Health and Safety at Work Act 1974 (HASAWA), which came into force from 1975 onwards, forced chemical plant owners to adopt more effective analysis techniques to identify hazards and develop measures to limit the consequences of accidents. The use of techniques such as hazard and operability studies (HAZOPS) and the probabilistic safety assessment of protective systems became common. The use of human error analysis techniques such as those described in Part I of this book is, however, still not widespread within the chemical industry.

This chapter also examines two accidents occurring in the public transportation industry, involving management failures of two very different kinds. One of these was the capsize of the cross-Channel ferry 'Herald of Free Enterprise' in 1987 which led to 192 fatalities. Here, the accident inquiry pointed unequivocally to serious failures at the highest levels of management of the company that owned and operated the vessel. The final case study in this chapter is more topical, if not controversial, and concerns failures at the organizational level within the UK privatized railway industry. It is suggested that the root cause of a number of recent railway accidents may be found in the way the privatized railway is organized. The faults then become attributable to policies of the government of the day put into effect by civil servants in the form of the privatization proposals.

It must be stressed that many of the human error analysis techniques described in Part I of this book are not easily applied to the categories of error addressed in this chapter. Nevertheless, by studying and learning from such accidents it is possible to identify potential flaws in the safety culture and management structures of organizations. The study of accidents caused by management and organizational failures reveals that the root causes were endemic long before an accident occurred; they are latent errors. If latent errors can be identified then the possibility exists that they can be corrected before they result in an accident. For this reason the reports of accidents such as those described below should be compulsory reading for senior managers and directors of companies so that they may recognize similar problems in their own organizations. Unfortunately, this recognition is often prevented by a tendency to assume that a similar accident could not happen in their company. It is hoped that the case studies described below will improve the understanding of how management errors can occur within an organization and demonstrate how the effect of these can be propagated downward in dangerous ways.

7.2 The Flixborough chemical plant disaster

7.2.1 Introduction

The Flixborough chemical plant disaster was the largest peacetime explosion ever to occur in the UK. It happened on Saturday, 1 June 1974 at the Flixborough chemical plant owned by Nypro (UK) Limited. It resulted in the deaths of 28 workers on the site and caused widespread damage to property within a 6 miles radius around the plant. The chemical plant at Flixborough was located in North Lincolnshire in an area that was predominantly agricultural and with a very low population density. In the 1960s, Dutch State Mines (DSM) of Holland developed a new process for the production of caprolactam, the monomer for Nylon 6. The popularity of nylon products at the time meant that in 1967 a 20,000 tonnes per year caprolactam plant was built by DSM at the Flixborough site using a process involving the hydrogenation of phenol. In this process, phenol was converted to cyclohexanone, one of the main intermediate chemicals in the production of caprolactam. In 1972, a major extension was completed bringing site capacity to 70,000 tonnes of the product per year, but using a more economic process. The new process was based on the oxidation of cyclohexane. It posed a much greater hazard than the phenol process and involved the recirculation of a large inventory of cyclohexane into which a stream of air was admitted. Cyclohexane is a highly volatile and flammable hydrocarbon liquid, and in this process is held under pressure

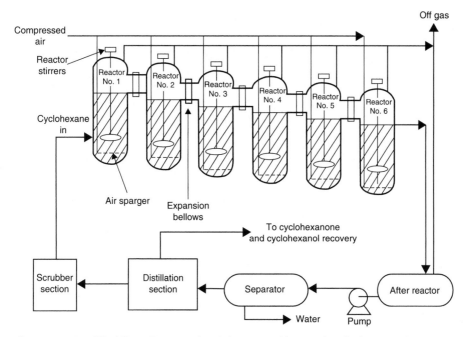

Figure 7.1 Simplified flow diagram of cyclohexane oxidation plant before March 1974.

above its atmospheric boiling point. This gave the potential for the large release of explosive vapour which ultimately led to the Flixborough disaster.

7.2.2 Description of the cyclohexane process

The process operates by injecting a stream of compressed air into liquid cyclohexane at a working pressure of about 9 bar (1 bar is approximately the same as atmospheric pressure) and a temperature of 155°C. It produces a mixture of two chemical compounds, cyclohexanone and cyclohexanol. It is a very inefficient process and it is necessary to recirculate the cyclohexane continuously through a train of six large stainless steel reactors. A simplified arrangement of the cyclohexane oxidation process is shown in Figure 7.1.

The cyclohexane flows through the six reactors by gravity into an after-reactor to complete the conversion process. From here it is pumped via a separator to a distillation process for recovery of the cyclohexanone product. Due to the extremely low conversion rate of cyclohexane to cyclohexanone it is necessary for the system to operate with a large inventory of cyclohexane. In order to accommodate expansion and contraction of the reactors and pipework at the temperature of operation, the 28 inch diameter overflow pipes joining each reactor are fitted with stainless steel bellows.

7.2.3 Events leading to the accident

7.2.3.1 *Technical problems and the miners' overtime ban of 1973*

Throughout 1973, the new caprolactam plant at Flixborough experienced serious technical difficulties and was unable to meet its assigned production targets. In November 1973 the technical problems were exacerbated by a miners' overtime ban which resulted in the government declaring a state of emergency and passing legislation to restrict the use of electricity by industry to 3 days a week. Since it was not possible to operate the process on this basis, it was decided to utilize existing emergency power generation on-site. However, management made arrangements to shut down unnecessary electricity consumers to enable production to be maintained as power generation capacity was insufficient to meet the full needs of the process.

One of the major electricity users was the six electrically driven mechanical stirrers in the cyclohexane reactors. The primary purpose of these stirrers was to disperse the compressed air that was injected into each reactor via a 'sparger' (a length of perforated pipe located at the bottom of the reactor). The stirrers also ensured that any droplets of water formed within the reactor system, especially at start-up, were dispersed into the cyclohexane. This was to prevent the droplets coalescing and settling as a layer at the bottom of the reactors. The droplets were carried forward in the recirculating stream of cyclohexane and removed in a separator immediately following the after-reactor.

7.2.3.2 *The No. 5 reactor problem*

By January 1974, the miners' overtime ban was resolved and normal electricity supplies had been resumed. On re-commissioning the reactor stirrers it was found that the drive mechanism for the stirrer in the No. 4 reactor had been subject to severe mechanical damage. No reason was found for this. It was therefore decided to continue to operate the plant with the No. 4 reactor stirrer shutdown. By this time, the earlier production problems had been overcome and it was possible to increase production. However, very soon after the plant had achieved its normal yield, a new problem arose which threatened the production gains already made.

The cyclohexane reactors were mild steel pressure vessels fitted with an inner stainless steel lining to resist the corrosive effects of organic acids produced in the oxidation process. In March 1974, cyclohexane was found to be leaking from a 6 feet long vertical crack in the mild steel shell of the No. 5 reactor. This indicated that the inner stainless steel lining was also defective. In normal times it might have been prudent to shut down the process while the reason for this serious structural failure was investigated, in case other reactors might have been affected. These were, after all, pressure vessels containing a large inventory of hydrocarbon liquid at above its atmospheric boiling point. Any sudden failure of a reactor had the potential for a major release of inflammable vapour and liquid particulate, not dissimilar to the effect of removing the cork from a champagne bottle. Due to the technical problems experienced earlier and the effects of the 3-day week, the owners of the plant were keen to make up the lost

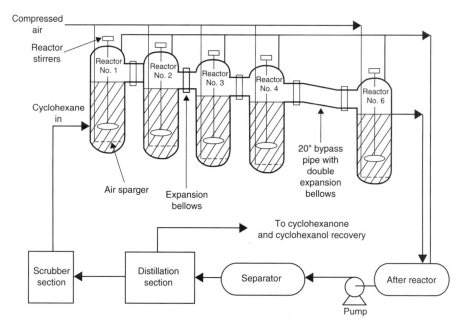

Figure 7.2 Simplified flow diagram of cyclohexane oxidation plant after March 1974.

production. It was therefore decided to remove No. 5 reactor for inspection and continue operations with the remaining five reactors.

7.2.3.3 *The installation of the 20 inch bypass pipe*

In order to continue operation it was decided to bypass the faulty No. 5 reactor using 20 inch diameter pipe. This pipe connected together the existing 28 inch bellows on the outlet of reactor No. 4 and the inlet of reactor No. 6. Due to the difference in level between the inlet and outlet pipes, the bypass was in the shape of a dog-leg with mitred joints. Unfortunately at the time the decision was made, the company did not have a properly qualified mechanical engineer on site to oversee the design and construction. The method of building the bypass involved laying the pipes on a full-scale template drawn in chalk on the workshop floor, so that they could be cut and welded together. In order to avoid excessive strain being placed on the double bellows the assembly was supported by scaffolding. No hydraulic pressure testing of the pipe, as required by British Standards, was carried out, except for a leakage test using compressed air. The arrangement of the bypass pipe installed in place of the No. 5 reactor is shown in Figure 7.2.

7.2.3.4 *Resumption of production*

The cyclohexane oxidation plant was restarted and operated normally, with occasional stoppages, up until the afternoon of Saturday, 1 June 1974. On the previous day the plant had been shut down for minor repairs associated with a leak. In the early hours of 1 June the plant was in the process of being restarted following the shutdown. The start-up involved charging the system with liquid cyclohexane to normal level and then recirculating this liquid through a heat exchanger to raise the temperature. The pressure in the system was maintained with nitrogen initially at about 4 bar until the heating process began to raise the pressure due to evaporation of cyclohexane vapour. The pressure was then allowed to rise to about 8 or 9 bar, venting off nitrogen to relieve any excess pressure. The temperature in the reactors by then was about 150°C.

On 1 June this procedure was followed except it was noted by the morning shift that by 06.00 hours the pressure had reached 8.5 bar even though the temperature in the No. 1 reactor had only reached 110°C. It was not realized at the time that this discrepancy might have indicated the presence of water in the system. By 15.00 hours the pressure was 7.8 bar and the temperature in the reactors was about 155°C. No record exists of the process conditions after this time. The start-up continued until, at about 16.50 hours, a shift chemist working in the laboratory close to the cyclohexane reactors, heard the sound of escaping gas and saw a haze typically associated with a hydrocarbon vapour cloud. He called for an immediate evacuation of the building.

7.2.4 The accident

At about 16.53 hours on 1 June 1974 a massive aerial explosion occurred with a force later estimated (from ionospheric readings) to be about 15–45 tonnes of TNT equivalent. Following the explosion, fierce fires broke out in many parts of the plant and burned out of control for 24 hours. The explosion was heard up to 30 miles away and damage was sustained to property over a radius of about 6 miles around the plant. Twenty-eight plant workers were killed with no survivors from the control room where the plant was being operated. In addition all records and charts for the start-up were destroyed, making it difficult for the investigators to later piece together the precise sequence of events in the final hours before the explosion.

Following the explosion, the 20 inch bypass assembly was found jack-knifed beneath the reactors with the bellows ruptured at both ends. The consequences of the disaster would have been far worse had it occurred on a normal working day rather than on a Saturday, since there would have been about 100 office staff working in the headquarters building about 100 metres from the plant.

7.2.5 The public inquiry

Following the disaster, it was announced that a public inquiry would be conducted under the chairmanship of Roger Parker QC, supported by a panel of experts. The terms of reference were to establish the causes and circumstances of the disaster and to identify any lessons that could be learned. The Court of Inquiry commenced on 2 July 1974 and lasted for a total of 70 days with evidence being received from 173 witnesses. The work of recovering, identifying and examining the wreckage, and of conducting relevant tests was undertaken by the Safety in Mines Research Establishment, Sheffield. In addition consultants were appointed to advise the various parties concerned.

7.2.5.1 *Conclusions of the inquiry*

The main conclusion of the public inquiry (Roger Jocelyn Parker QC, 1975) was that the immediate cause of the main explosion was the rupture of the 20 inch bypass assembly between the No. 4 and No. 6 reactor. As to how this occurred, two main theories emerged from the investigations.

7.2.5.2 *The 20 inch pipe theory*

The simpler theory held that the 20 inch bypass assembly failed due to its unsatisfactory design features. However, this could not really explain how the assembly had survived 2 months of normal operation. In order to determine whether on the day of the explosion unusual conditions on the plant might have caused the bypass to fail, a number of independent pressure tests were commissioned using three full-scale simulation rigs at different locations. The normal working pressure of the reactors was about 8 bar but it was established that there was a practice during start-up to allow the pressure to build up beyond the working pressure to about 9 bar. The safety valves for the system, which would have relieved any excess pressure, were set to discharge at a pressure of 11 bar, this being the maximum achievable pressure. During the simulations it was found that at pressures in excess of the safety valve pressure of 11 bar, the replicated assembly was subject to a squirming motion which distorted the bellows. However, even when the assembly squirmed in this way it was not possible to cause a rupture until the pressure was increased to about 14.5 bar, a pressure not achievable in the Flixborough reactors because of the safety valves.

The Inquiry concluded, nevertheless, that a rupture of the 20 inch bypass assembly must have taken place at pressures and temperatures that could 'reasonably be expected to have occurred during the last shift'. The report conceded that this conclusion was 'a probability albeit one which would readily be displaced if some greater probability to account for the rupture could be found'. It was reached in the face of the evidence that none of the three independent simulation tests had been able to replicate the failure at any pressure achievable within the system.

7.2.5.3 *The 8 inch hypothesis*

The only other theory put forward by the Inquiry to explain the disaster, was known as 'the 8 inch hypothesis'. This referred to a 50 inch split found in an 8 inch line connected to the separator below the bypass. It was speculated that this failure might have led to a smaller explosion precipitating the failure of the main 20 inch bypass. Metallurgical examination of the 8 inch pipe showed that a process known as zinc embrittlement had caused the split. It was suggested that a small lagging fire had occurred at a leaking flange causing zinc from a galvanized walkway above the fire to drip onto the 8 inch pipe. This in turn resulted in a brittle failure of the pipe, causing a vapour release that ignited, exploded and disturbed the 20 inch bypass pipe causing it to fail. About half of the Inquiry Report is devoted to a discussion of this two-stage theory which was finally dismissed as being too improbable. No other theories were considered by the Inquiry at the time to explain the reasons for the failure of the 20 inch bypass pipe.

7.2.5.4 *The water theory*

One other theory was put forward but for reasons that are not recorded was not considered by the Inquiry. However, much of the scientific work to underpin the theory did not take place until after the Inquiry had closed. The theory examined the effects of not operating the No. 4 reactor stirrer during the start-up at a time when water might have been present in the system. It later became known as the 'water theory' and came to be highly significant in providing a more probable explanation for what remains an unexplained disaster.

Cyclohexane and water are normally immiscible liquids and will separate into two layers after being mixed together. The water theory discovered that at the interface between these two layers, a liquid known as an azeotrope forms due to the limited solubility of water in cyclohexane. This azeotrope has a lower boiling point than either water or cyclohexane. Thus, if a layer of water is allowed to settle in a pool below a layer of hot dry cyclohexane in the reactors, an unstable interfacial layer may form which under certain conditions can boil and erupt violently ejecting cyclohexane and superheated water from the reactor.

Under normal operating conditions, it is impossible for a water layer to form due to the effect of the distribution of air into the bottom of the reactor which disperses any water droplets preventing them from settling. During start-up, however, the air to the reactors is shut off until the temperature of the cyclohexane approaches the operating temperature. If the stirrers are running during start-up, then water cannot form a layer. However if a stirrer is stopped, as the No. 4 stirrer had been stopped on 1 June, due to the failure of the drive mechanism, a layer of water could form, together with the unstable azeotrope. As the temperature of the reactor contents increases during start-up, the boiling point of the azeotrope is reached with the possibility of a sudden violent eruption from the reactor. This eruption is accompanied by the ejection of slugs of liquid reactant. These slugs could have exerted high mechanical forces on the bypass assembly, which was only loosely supported by scaffolding. This in turn could cause the bypass assembly to fail without the need to develop a high static pressure in the reactors.

7.2.6 Alternative event sequence

Although the water theory was not considered by the Inquiry, it provides a more credible explanation for the cause of the disaster than the other theories. Not only does it provide an explanation for the failure of the 20 inch bypass pipe on the day of the disaster, it also provides an explanation for the whole sequence of events that took place over the preceding 6 months. This alternative event sequence was as follows:

1. the unexplained failure of the drive mechanism for the No. 4 reactor;
2. the development of a crack in the lining and shell of the No. 5 reactor;
3. the failure of the 20 inch bypass assembly.

Any or all these failures could have been caused by the violent eruption of the reactor contents due to the presence of water. The Inquiry does not appear to have considered that the three events could be linked by a common cause.

No cause was assigned to the unexplained failure of the drive mechanism for the No. 4 reactor. It was presumably thought not to be relevant since no mention of it occurs in the Inquiry Report.

The crack in the shell of the No. 5 reactor was attributed to a mechanism known as *nitrate stress corrosion*. This theory was developed by the plant owners as an explanation of the crack, but subsequently fell into disrepute.

The failure of the bypass and the direct cause of the disaster itself were concluded by the Inquiry to be due to the reactors being over-pressurized. This, by implication, must have been caused by human error on the part of the plant operators, none of whom survived the disaster.

Perhaps the greatest failing of the Flixborough Inquiry was that it did not take account of all the events that took place during the 6 months preceding the disaster. Its investigations did not go far enough back in time. The whole technical issue of non-operation of the reactor stirrers was ignored by the Inquiry for reasons that have never been explained.

7.2.7 Epilogue

In recent years there has been a resurgence of interest in the causes of the Flixborough disaster and this has resulted in the UK Health and Safety Executive (HSE) commissioning a series of laboratory experiments to investigate the 'water theory'. Early results suggest that the presence of water may indeed have been significant. Larger scale tests by HSE are in progress at the time of writing; hence the issue currently remains unresolved. It is possible that the main conclusion of the inquiry, that the disaster was caused by the failure of an inadequately designed temporary bypass pipe exacerbated by an error on the part of the plant operators, may yet be overturned.

7.2.8 Conclusions

7.2.8.1 *Human error analysis*

This section summarizes the main human error causes lying behind the Flixborough disaster in terms of its direct, root and contributory causes. Table 7.1 tabulates these causes against the different types of error that occurred.

7.2.8.1.1 Direct cause

The direct cause of the disaster was the failure of the 20 inch bypass pipe between reactors No. 4 and No. 6. This caused a massive release of inflammable cyclohexane vapour which ignited, causing an unconfined vapour cloud explosion.

7.2.8.1.2 Root causes

The root cause of the disaster lies in the reasons why:

- a badly designed 20 inch bypass pipe was installed rather than the reasons for the crack in the No. 5 reactor being investigated?
- having installed the bypass, what caused it to fail on 1 June?

Table 7.1 Human error analysis of the Flixborough disaster

Cause	Description	Type of error or other failure	Systemic cause or preventive action
Direct cause	Failure of the 20 inch bypass pipe causing a massive release of cyclohexane vapour to the atmosphere	Operator error (implied)	Since the inquiry was unable to establish a definitive cause (only the one with the highest probability), then by implication, the direct cause had to be over-pressurization of the system by the operators at start-up.
Root cause	The cause of the 20 inch bypass failure was either: Over-pressurization during start-up or,	Management error	Inadequate design and support of the bypass pipe in a dog-leg configuration due to lack of mechanical engineering expertise.
	Mechanical displacement of the bypass pipe by slugs of liquid carried forward from reactor No. 4	Design error	A failure on the part of the designers to recognize or communicate the consequences of not operating the reactor stirrers at start-up.
Contributory causes	Failure by local management to understand the hazards of the cyclohexane process	Management error	This knowledge might have prevented the expedient construction of the 20 inch bypass pipe, due to the perceived hazards of a large unconfined vapour release.
	No qualified mechanical engineer on-site	Management error	Changes to a design should be overseen and authorized by properly qualified personnel.

The reasons for installing the bypass were simple expediency made under commercial pressure. Commercial interests took precedence over safety. The reason for the failure of the bypass is another matter entirely. Even 25 years later this still remains a controversial issue, with more than one theory competing for attention. If the Inquiry findings are correct, then the root cause of the disaster is indeed a decision-making error by local management. However, if the 'water theory' proves to be the more likely cause, then the disaster may be attributable to a design error, or at least a failure on the part of the designers to understand or, if they knew about it, to communicate to local management the catastrophic consequences of allowing a water layer to build up in the reactors.

7.2.8.1.3 Contributory causes

A contributory cause of the disaster was a failure on the part of management to recognize the hazards of the process. If the potential consequences of a release of hundreds of tonnes of cyclohexane vapour to atmosphere had been understood, then it is possible that the reasons for the crack in the No. 5 reactor would have been properly investigated. Another reason for the failure to investigate was the lack of a properly qualified mechanical engineer on the management team with sufficient authority to override the decision.

7.3 The capsize of the Herald of Free Enterprise

7.3.1 The accident

The 'Herald of Free Enterprise' was a roll-on/roll-off (RORO) car ferry operating between the British Channel ports and continental Europe. The vessel was operated by Townsend Thoresen and had a maximum carrying capacity of 1400 passengers. On the evening of 16 March 1987 the ferry left the Belgian port of Zeebrugge with 80 crew and 459 passengers on board on what should have been a routine crossing to Dover. Within about 20 minutes of departure water began entering the car deck as a result of a failure to close the bow doors prior to the vessel leaving port. After passing through the inner breakwater of the harbour, the ferry accelerated to about 15–20 knots. Here the waves were of sufficient height to flood the car deck. Water entered through the open bow doors at a rate of around 200 tonnes/minute causing the ferry to list to port and finally to roll over onto a sand bank near the harbour entrance. Due to the speed at which the disaster developed there was little chance for preparation, no warnings or announcements were made and passengers were left mainly to their own devices to escape from the half submerged ferry. The disaster resulted in the deaths of 192 people, the most serious accident involving a UK passenger vessel for many decades. It was already known that RORO ferries of this design were highly susceptible to capsizing with only a few inches depth of water present on the car deck. In fact only 5 years earlier the same company had experienced the loss of its ferry European Gateway when it capsized in 4 minutes with the loss of six lives at the approach to the port of Harwich following a collision.

7.3.2 The inquiry

7.3.2.1 *Scope*

A public inquiry was held into the disaster in 1987 under the Wreck Commissioner, Justice Sheen (1987). The Court of Inquiry was held under the Merchant Shipping Act 1984. Such a court has full powers of investigation, the power to suspend or remove a Merchant Officer's Certificate of Competency and is also able to determine who should pay the cost of the investigation. Curiously, however, the court does not have any power to make binding recommendations upon any of the parties, relating for instance to design or operational issues. This is particularly relevant to this accident, because the consequences were exacerbated by a basic design flaw that is common to most currently operating RORO ferries. It would be quite possible to include this accident in the following chapter, which concerns human errors in design. However, it has been included here because the principal cause discovered by the public inquiry was serious deficiencies within the culture of the owner's organization and errors made by management from board level downwards.

7.3.2.2 *Inquiry findings*
7.3.2.2.1 Direct cause
The Sheen Inquiry found that the direct cause of the accident was the absence of the Assistant Bosun of the vessel from his post at the time of sailing from Zeebrugge. The Assistant Bosun was the principal person responsible for ensuring that the bow doors of the vessel were closed before it left port. It was established that he was asleep in his cabin at the time of departure. It was therefore an error of omission. This prompted the court to examine working practices in the company, particularly the long hours which staff were required to work without rest periods. The Inquiry thought this was a contributory factor. However, a more important factor was the absence of any engineered means of error recovery if the person responsible for closing the bow doors failed to do so. In particular, there was no provision to indicate to the captain or senior officer on the bridge whether the bow doors were open or shut; the position of the doors was not visible from the bridge. The importance of engineered recovery mechanisms to limit the effects of catastrophic single human errors has already been emphasized in Part I of this book, and this is an extremely good example of the lack of such a mechanism.

In the absence of any second line of defence against error, an informal but dangerous system had been adopted. The captain operated on the simple assumption that if he heard nothing from the Assistant Bosun then he would assume that the bow doors had been closed. No further independent checks were carried out. The result was that on this occasion the captain ordered the vessel to sea with the bow doors still open.

7.3.2.2.2 Root causes – culture and management of the organization
The Sheen Inquiry allocated the highest degree of responsibility for the accident to senior management. The responsibility extended as high as Board of Director level. The problem was principally lack of communication between operational management, in particular between the Masters of vessels, who had serious concerns about safety, and higher levels of management onshore. The communication tended to be one way – upwards. The safety concerns of the Masters had been expressed in countless memoranda to top management but these were consistently ignored. Requests for safety improvements were seen at best as involving unwarranted expenditure and at worst were regarded as criticism of the company. This resulted in a stand-off between offshore and onshore managers with the latter adopting a defensive posture and a refusal to act. Very few meetings ever took place. As an example of such lack of communication, one of the senior Masters had reported concerns about overloading of vessels and the fact that when vessels were trimmed waves would be lapping over the bow doors. The senior manager who failed to act, held the view that if the matter was really serious, the senior Master would have come to see him and 'banged on his desk' rather than send a memo. Overloading and poor trimming of the ship was a significant factor in the 'Herald of Free Enterprise' disaster. The Sheen Inquiry concluded that:

> ... *a full investigation into the circumstances of the disaster leads inexorably to the conclusion that the underlying or cardinal faults lay higher up in the Company. The Board of Directors did not appreciate their responsibility for the safe management of their ships. They did not apply their minds to the question: What*

orders should be given for the safety of our ships? The directors did not have any proper comprehension of what their duties were. There appears to have been a lack of thought about the way in which the Herald *ought to have been organized for the Dover–Zeebrugge run. All concerned in management, from the members of the Board of Directors down to the junior superintendents, were guilty of fault in that all must be regarded as sharing responsibility for the failure of management. From top to bottom the body corporate was infected with the disease of sloppiness ... it reveals a staggering complacency.*

The predictable reaction of senior management to the failure to close the bow doors prior to departure was to assign blame to the Assistant Bosun for being asleep at his post. It exhibited a lack of understanding by management that it was they who were responsible for providing systems to guard against the consequences of simple operational errors that would inevitably happen from time to time. The possibility that the Assistant Bosun might not be present at his post on departure, and its consequences, had already been foreseen by one of the Masters and raised with senior management in the form of a request for a bow door position indicator to be provided on the bridge. The reaction of senior management was to dismiss this request, on the justification that the provision of expensive equipment to protect against an employee failing to undertake his duties was frivolous in the extreme. Failures within the organization and management were the principal root cause of the accident. There were a number of other contributory causes which are briefly discussed below.

7.3.2.2.3 Contributory causes
Operational problems
An important contributory cause of the disaster was a failure properly to trim the vessel before leaving port. This caused the vessel to be down at the bow when it departed, making it more susceptible to the entry of water on to the car deck. One of the reasons for the poor trimming was that it was necessary to fill the ship's bow ballast tanks for the ferry to be loaded and unloaded at Zeebrugge so that the car deck was level with the ramp. When at sea, with the ship's bow low in the water, waves would come three-quarters of the way up the bow doors, posing a severe hazard to the ship. This matter was raised with senior management by ferry captains some 6 months before the disaster, but no action was taken. There were a number of other contributory causes identified by the Inquiry. One of these was a failure by the company to have a system in place so that ferry captains would know how many passengers were on board at departure. This was important in the event of an accident or emergency requiring evacuation of passengers. Following the accident at Zeebrugge, there was no definitive passenger list available to indicate the number of passengers on board.

The design flaw
Although it is more appropriate to discuss design errors in Chapter 8, it is worth highlighting here a basic design fault which still exists in many existing RORO ferries operating between British ports and the continent. For many years before this accident naval architects had been questioning the design of RORO ferries with regard to their inherent

instability compared with conventional vessels. Tests on a ship model carried out by the Danish Maritime Institute after the accident, found that this design of ferry took less than 30 minutes to capsize once water had started to enter the car deck. In fact, the 'Herald of Free Enterprise' turned over in just 45 seconds. Under the Safety of Life at Sea (SOLAS) design rules (an international convention administered by the International Maritime Organization or IMO) in force when the vessel was designed and built, RORO ferries were required to have at least 76 millimetre freeboard. In this context freeboard refers to the distance between the vehicle deck and sea level. It was quickly recognized following the 'Herald of Free Enterprise' disaster, that adequate freeboard is crucial in preventing a dangerous ingress of water on to the car deck. If there is damage to the bow doors of a ferry, or if the doors are left open, then where the wave height is greater than the freeboard it is largely a matter of chance how much water will enter and whether a capsize will occur. Since 1990, the required freeboard has been increased under a new SOLAS rule to 1.25 metres and all new RORO ferries are required to comply with this. Despite pressure upon the IMO from the British Government these rules have not been applied retrospectively to the large number of existing ferries which continue to operate at high frequency across the English Channel. It has been estimated by Lloyd's Register that the 'appropriate ingredients' for a repetition of the Zeebrugge disaster occur on average once every 5 years, usually as a result of a ship collision.

Tests carried out since the disaster on models of RORO ferries have identified that relatively small amounts of water present on the vehicle deck are sufficient to destabilize a high sided ferry leading to a rapid capsize. Once water begins to accumulate on the car deck, it will naturally flood to one side or other of the vessel producing a list to starboard or port. As the amount of water increases, so does the list until the vessel finally capsizes. One way of avoiding this problem is to provide watertight bulkheads or longitudinal dividers which subdivide the vehicle deck and prevent all the water flooding across to one side of the vessel thus reducing the likelihood of a capsize. Unfortunately, these bulkheads also impede vehicle loading on to the ferry and the industry regards this solution as impractical.

The design and instability issue relating to RORO ferries was hardly raised at the Public Inquiry into the 'Herald of Free Enterprise' disaster, even though it undoubtedly contributed to the large loss of life. The Inquiry was not empowered to issue design recommendations. One is reminded of the failure of the Flixborough Inquiry to address design issues some 20 years earlier (see above). There is often reluctance on the part of management and industry seriously to question design criteria for equipment, where changes to these criteria would prove expensive or be perceived as impractical. The result is an over-emphasis on the direct causes of the accident, often associated with human error at the operational level, rather than the root causes associated with the design.

7.3.3 Conclusions

7.3.3.1 *Human error analysis*

Table 7.2 summarizes the main human error causes of the accident as described above.

Table 7.2 Human error analysis of the 'Herald of Free Enterprise' disaster

Cause	Description	Type of error or other failure	Systemic cause or preventive action
Direct cause	Failure of Assistant Bosun to close bow doors before departure	Active error	Excessive work hours. Unacceptable system assuming bow doors were closed if no message was received. Engineered system of error recovery was required – bow door position indicator on bridge.
Root cause	Lack of effective communication between Ship's Masters and higher levels of management onshore	Management error	Requests for safety improvements by Ship's Masters were consistently ignored by onshore management.
	Onshore management and company directors did not have any proper comprehension of what their duties and responsibilities were with regard to safety	Cultural error	The culture of the company did not give safety any priority over commercial considerations, such as maintaining a high turn around of ferries, and safety expenditure was regarded as an unnecessary cost.
Contributory causes	Instability of RORO ferry design	Design error	Industry had failed to take cognisance of the proven instability of this type of ferry with only a few inches of water entering the car deck. Refusal to consider retrospective implementation of improved standards for freeboard.

7.4 Privatization of the railways

7.4.1 Introduction

The privatization of the UK railway network took place in 1997 and introduced fundamental organizational and management changes having a potential impact upon railway safety. Since privatization, a strong public perception has developed that the railways are less safe than they were when nationalized. This perception has been mainly brought about as a result of a series of multiple fatality train accidents. This perception is supported by accident statistics presented below. Of the accidents that have taken place since privatization, those which can be most closely linked to the privatization process were caused by latent maintenance errors. A number of these accidents, in particular the accidents at Hatfield and Potters Bar, are presented as separate case studies in Chapter 9. Since privatization there have also been a number of accidents caused by active errors, in particular by train drivers passing signals at danger. The more recent of these accidents at Ladbroke Grove and Southall in London are presented as case studies in Chapter 10. In this chapter the degree to which changes in the

Table 7.3 Fatal accidents and fatalities on the UK national railway system 1967 to June 2002

Period	Train-kilometres (billion)	Number of fatal accidents	Number of fatalities	Accidents per billion train-kilometres	Fatalities per accident	Fatalities per billion train-kilometres
1967–1971	2.25	26	119	11.6	4.6	52.9
1972–1976	2.18	16	42	7.3	2.6	19.3
1977–1981	2.13	10	28	4.7	2.8	13.1
1982–1986	1.99	12	36	6.0	3.0	18.1
1987–1991	2.15	10	57	4.7	5.7	26.5
1992–1996	2.13	6	11	2.8	1.8	5.2
1997–June 2002	2.71	5	58	1.8	11.6	21.4
1967–June 2002	**15.54**	**85**	**351**	**5.5**	**4.1**	**22.6**

Source: Andrew Evans, Professor of Transport Safety, University College, London (Evans, 2002).

management and organization of the railways may have contributed to these accidents is considered. However, prior to an examination of these issues, it is instructive to examine railway accident statistics both pre- and post-privatization.

7.4.2 Rail safety statistics before and after privatization

Table 7.3 presents railway accident statistics over 4–5 year periods from 1967 to June 2002. Of particular interest for this chapter are the 5-year periods before and after 1997, the year of privatization. This is in order that a valid safety comparison can be made.

The accidents include fatal collisions, derailments and overruns of red signals all involving multiple fatalities but exclude, for instance, single fatality accidents caused by people falling from trains, station platforms, pedestrian bridges, etc. The periods of most interest are those pre- and post- railway privatization, namely 1992–1996 and 1997 to June 2002, respectively. It can be seen that while travelled distance (train-kilometres) has increased slightly from 2.13 to 2.71 billion in the two periods, respectively, the number of fatal accidents has decreased slightly from 6 to 5. If the number of fatal accidents per billion train-kilometres is calculated it is seen that the fatal accident rate has decreased from 2.8 to 1.8. At first sight, this would seem to disprove the popular perception that rail travel is less safe under privatization. However, it is also seen that the number of fatalities in each of these periods has increased from 11, before privatization, to 58 after privatization. This has occurred because of a greater number of average fatalities per accident, this having increased from 1.8 to 11.6. This large increase indicates that although the number of accidents has decreased very slightly, the severity of the accidents has worsened in terms of the number of fatalities caused.

The statistic which is of most interest to the rail traveller is not so much the fatal accident rate but the chance of being killed during a journey to work. This statistic, which can be expressed in a variety of ways, is the one that would normally be used to estimate the numerical risk of rail travel. It would be easily comparable for instance with the risks of travelling by air or road which are calculated in much the same way.

From Table 7.3 it is calculated that the risk of fatality per billion kilometres travelled by rail has increased from 5.2 deaths per billion kilometres travelled before privatization, to 21.4 deaths per billion kilometres travelled after privatization, over the respective periods. Although the risks to an individual passenger are still extremely small compared, for instance, with driving a car, this roughly represents a fourfold increase in risk and tends to support the public perception that rail safety has worsened since privatization. There is of course always an element of uncertainty in statistics interpolated over short periods, but this comparison is still rather alarming. The fourfold increase has occurred over a period where crashworthiness of rolling stock has improved and one would expect to see an increase in survivability of accidents, not the converse. On the other hand, most of the high fatality crashes that have occurred involved high-speed trains, and this has contributed to the increased severity of the accidents. There is therefore an element of chance involved.

Most of the multiple fatality accidents which have occurred since privatization have either involved high-speed derailments due to track faults (Hatfield and Potters Bar) associated with maintenance of the infrastructure or signals passed at danger (Southall and Ladbroke Grove) associated with driver error. Although it is more difficult to attribute accidents due to driver error to management factors related to privatization, it is easier to make a connection with accidents due to track related maintenance failures. The latter are more likely to be linked to changes in organizational arrangements following privatization. This link is examined more closely in this chapter and in Chapter 9.

7.4.3 A brief history of railway safety management

Prior to 1948, the railway industry was 'privatized' in that the rail network was owned and operated by private railway companies. The main difference between the privatized railway pre-1948 and the railway since 1997 was in the responsibility for trains and infrastructure. Prior to 1948, the railway companies were 'regional' in nature and each company was responsible for both the infrastructure and the trains in the region in which it operated. They operated what has become known today as a 'vertically integrated railway', where a company owns and operates both rolling stock and infrastructure which includes track, signalling and stations. For most of the 20th century, safety was regulated by the 1889 Regulation of Railways Act, which remained in force up until privatization in 1997. From 1923, the host of private railways that operated in the UK were amalgamated by government decree into what was known as the 'big four': The London, Midland and Scottish Railway (LMS), the London and North Eastern Railway (LNER), The Great Western Railway (GWR) and the Southern Railway. This was the situation prior to nationalization in 1948 when the whole network was brought into public ownership.

An advantage of the newly nationalized railway was that, although it may have been a rather monolithic structure and, therefore, subject to a degree of bureaucracy and inefficiency, at least the whole industry had a common set of objectives. It was primarily service driven as opposed to profit driven, which was the situation in the previously 'privatized' railway. However, before and after nationalization, rail safety was not seen as a separate objective in its own right, but was integrated within a set of rules (known as the Rule Book) that governed operations. Railway safety as an objective in its own right suddenly became a matter of public concern following the Clapham Junction accident in 1989. The accident at Clapham Junction was a watershed for the railway industry's own perception of safety. Following Clapham, safety became fully acknowledged as a specific objective. Prior to that it was generally accepted that the Rule Book approach and safety were synonymous. This was to some extent true, and in one sense the ideal way to run a safe railway is seamlessly to integrate safety and day-to-day operation.

Privatization of a whole tranche of public utilities took place in the 1980s under the Conservative government led by Margaret Thatcher, culminating in what was perhaps the greatest challenge of all, the nationalized railway. In 1997, the government's proposals for railway privatization resulted in the break-up of the railway into more than 100 private companies. These encompassed rail passenger and freight operations, maintenance companies for track and equipment and rolling stock leasing companies. It effectively ended the vertically integrated railway that had existed since the 19th century. A separate company, Railtrack plc, was formed and became the 'infrastructure controller', charged with ownership and operation of the track, stations and signalling for the whole network. Operation of rolling stock was now by 'Train Operating Companies' (TOCs) to whom franchises were let by the government on a regional and renewable basis. The TOCs paid track access charges to the infrastructure owner and controller. The trains themselves were leased from separate rolling stock companies (ROSCOs) who were responsible for repair and maintenance. Thus, responsibility for safety became fragmented. Communication and consultation about safety matters had to cross new organizational interfaces which did not exist in the nationalized railway. In addition the industry was governed by the profit motive rather than by public service, although the requirement to provide a public service was protected in the new legislation privatizing the railways.

7.4.4 The impact of privatization upon safety

7.4.4.1 *The safety case approach*

One of the major implications of privatization for safety was that the official safety regulator, Her Majesty's Railway Inspectorate (HMRI), took on a substantially different role. After the Clapham Junction accident in 1987, HMRI had been brought under the control of the Health and Safety Executive (HSE). In accordance with current HSE philosophy, a new approach was adopted. Instead of the traditional, prescriptive hands-on 'Rule Book' approach to safety administered by HMRI since the formation of the railways, a less prescriptive approach using Safety Cases was adopted to demonstrate safe operation.

Under the new approach, railway operators were required to demonstrate not so much compliance with prescriptive rules (rules which were vetted by the regulator and to which the rail operators had to conform) but to have in place their own set of rules which they must demonstrate were adequate to ensure safe operation, and were being followed. This demonstration was accomplished by means of the Safety Case, which became a requirement under new Railway (Safety Case) Regulations 1994 (HSE, 1994). After privatization, the infrastructure controller, at that time Railtrack, took responsibility for vetting Safety Cases submitted by the TOC. Before an operating company was allowed to bring its rolling stock on to the track, it had to demonstrate in a written Safety Case that all necessary operational safety provisions were in place. In turn, Railtrack had to produce their own Safety Case which was itself validated by the regulator, HMRI. This was in essence a three tier or cascaded system, where the regulator validates the infrastructure controller to validate the Safety Cases of others. Thus, although day-to-day operation was still governed by the Rule Book, regulation of this by HMRI now had a much more hands-off approach, and was also expected to take account of commercial realities.

7.4.4.2 *The dilution of knowledge*
In the period since HMRI came under the HSE its ethos has changed almost out of recognition. Much of the railway specific knowledge that existed in the mind of the traditional Railway Inspector became diluted as Health and Safety generalists from other branches of HSE were recruited to HMRI. As an example of this, the old Inspecting Division of HMRI, who used to go out and inspect the railway, investigate accidents and make recommendations, was absorbed into the Field Operations Directorate (FOD) of HSE which monitors health and safety on an industry wide basis.

Similarly, in managing the newly privatized railway, the owner of a TOC taking over a franchise did not necessarily need to possess detailed experience of running a railway. Those drawing up the privatization proposals assumed that a TOC would inherit from the previous owner the same middle management personnel, operating staff, management systems, Rule Books, operational methods and route knowledge which had applied previously. In theory, when a new company took over a British Rail (BR) regional service, such as the East Coast Main Line, it would inherit all the systems needed for safe running of the railway. The only visible change would occur at top management level.

There were a number of potential problems associated with this assumption. Firstly, because top management may not have a great deal of experience in running a railway, the company culture they brought to the operation might not be conducive to safe operation. Secondly, the main business motivation became profit rather than public service. Through legislation it was necessary for the government to closely regulate the businesses to ensure that safety was still a priority and that monopolistic abuse of fares and services did not take place. Safety was still monitored by HMRI, although its detailed knowledge of railway safety had become diluted before privatization by its integration into HSE. The protection of passenger interests was upheld by the office of the Rail Regulator, a completely separate role from that of the safety regulator, HMRI. One of the aims of the Rail Regulator was to ensure that conflicts between profit and public service did not arise and if they did were resolved quickly.

In spite of the measures to restrain the commercial instincts of the new companies, the main motivation for running a railway was still profit rather than public service. Much of the knowledge that existed within the industry in the minds of track maintenance engineers, signalling experts and other specialists, was quickly dispensed with. Almost immediately large-scale redundancies and pay-offs took place in order to reduce man-power costs. The effects of these and similar changes arising from privatization, together with their impact on safety, are discussed in more detail in Chapter 9 in relation to the way rail maintenance activities were managed.

7.4.4.3 *The loss of 'controlling mind'*

From the perspective of this book, industrial safety is a function of the way work is carried out and whether the existence of dangerous human errors can be detected and their effects adequately controlled. The main influence of privatization upon the way work is carried out has been the organizational changes stemming from the fragmen-tation of the industry. The loss of the vertically integrated railway resulted in the need for greatly improved communication and consultation across the increased number of operational boundaries. In this respect, one of the most important consequences of privatization was the loss of a single 'controlling mind' which, before privatization, could be found in the British Railways Board.

The 'controlling mind' was in charge of, responsible and accountable for, safety across the whole network. The vertically integrated railway enabled safety concepts formu-lated at the highest level to be filtered down and implemented consistently through the whole organization with the possibility of information from the lower levels being fed back to close the information loop. Following privatization, the integrated railway was deliberately broken down into a large number of competing separate businesses, each with its own management structure and independent objectives. Possibly the only common objective was now how much profit could be made for the shareholders, which in turn influenced the quoted share price of the company on the stock exchange. By the year 2002, the UK railway network was being operated by, it is estimated, 1600 separate companies, many of these companies being small subcontractors, resulting in further dilution of control and accountability. The number of interfaces between the parts of the organization responsible for different aspects of operation, increased in proportion. The problems brought about by the increased use of contractors are dis-cussed more fully in Chapter 9, Section 9.3, where the rail accident at Hatfield is used as a case study for maintenance errors.

7.4.5 Conclusion

The principal aim of privatization was to create competition and inject a new sense of commercial reality into a business that was making increasing demands for public subsidy and was in desperate need of massive capital investment. Previously this investment would have had come from the public purse. Competition was expected to produce new efficiencies and it was hoped that a better public service would eventu-ally emerge. Since privatization, an increase in train-kilometres travelled (see Table 7.3) has to some extent justified the original aspirations. There has been a substantial increase in both passenger and freight traffic across the network. However, this increase in traffic has also highlighted and exacerbated previously existing problems, not the least of which was the reluctance of previous governments to invest in modernization of the country's railway infrastructure and rolling stock. This was tragically illustrated by a serious fatal accident at Hatfield in Hertfordshire on 17 October 2000 due to track deficiencies. Following the accident, the problems with the track were further highlighted by a serious disruption of rail operations lasting many months while repairs were carried out over the entire network.

The Hatfield rail accident resulted in four fatalities and 70 injuries when a high-speed train was derailed due to a track failure. The accident is described in more detail in Chapter 9 and is one among a number of accidents resulting from inadequate maintenance of the network. The Hatfield accident was perhaps the most significant and brought to public attention how changes in railway management following privatization had resulted in new and unforeseen hazards to rail travellers. After Hatfield, it was acknowledged that Railtrack did not have a clear picture of the condition of the infrastructure, such that dangerous faults on the track were allowed to develop with the result that a serious accident became inevitable. Tragically, this situation was further underlined by an accident at Potters Bar some 2 years later (also described in Chapter 9) resulting in seven fatalities following a derailment at badly maintained points within a few miles of the location of the Hatfield accident. The root cause of both accidents is traceable back to human errors at government policy-making and Civil Service Executive level.

In conclusion, privatization was accompanied by a failure properly to understand the safety implications of introducing a new complex organizational structure into an industry, which itself was only just beginning to understand how safety should be managed on a modern railway. The lessons learned from the Clapham Junction accident by the British Railways Board some 10 years before privatization were still in the process of being implemented. By reference to Table 7.3 it is seen that the reduction in fatalities per billion train-kilometres travelled fell from 26.5 in the period 1987–1991 (following but including the Clapham Junction accident) to 5.2 in the period 1992–1996 (just before privatization). This is to some extent indicative of the progress that was made, although the Clapham Junction accident will have skewed the earlier statistics. As a result of privatization the tentative improvements in safety management were effectively put on hold, while the entire organization of the industry was re-engineered. The result was the loss of the 'controlling mind' and a major setback to rail safety.

References

Evans, A. (2002). Are train accident risks increasing? *Modern Railways Magazine*, August 2002, pp. 49–51.

Justice Sheen (1987). *The Merchant Shipping Act 1984: MV Herald of Free Enterprise. Report of Court No. 8074*, London: HMSO.

Roger Jocelyn Parker QC (1975). *The Flixborough Disaster. Report of the Court of Inquiry*, London: HMSO.

The Health and Safety Executive (2000). *Railways (Safety Case) Regulations 2000, Guidance on regulations L52*, Sudbury, Suffolk: HSE Books.

8

Design errors

8.1 Introduction

Design errors, together with maintenance errors which are considered in the next chapter, fall into the classification of latent errors described in Chapter 3. The effects of latent errors may not become obvious until an accident is initiated and this may be too late to prevent the accident turning into a disaster. The case studies in this chapter reflect the reality of this sad fact.

The first case study concerns an accident at a chemical plant at Grangemouth in Scotland. As an example of a latent error, it demonstrates the length of time that can elapse between the design error being made and the accident taking place. In this case, the period between the error and the accident was over 10 years. The direct cause of the accident was a simple active error by an operator. The possibility of such an error was not picked up at the design stage, mainly because the designers were under the impression that they had installed effective protective devices to prevent the consequences of the error being realized. However, a human error with a high probability coupled with a failure of the protective devices had the potential for a catastrophic failure of high-pressure equipment. This combination of human error and equipment failure was also not foreseen by the owners of the plant. In the event it was not necessary to wait for the protective equipment to fail for an accident to occur. The owners of the plant allowed the safety device to be bypassed and left it in this condition for a considerable period. It was only a matter of time before the now unprotected but inevitable human error would lead to catastrophe.

The second case study describes the sinking of the roll-on/roll-off (RORO) ferry 'Estonia' in the Baltic in 1994. The accident resulted in the highest loss of passenger life ever to occur during peacetime in a ship accident in Europe. On first sight, the accident has some resemblance to the Herald of Free Enterprise disaster described in Chapter 7. The fact that the ferry sank so quickly after the flooding of the car deck was largely a result of the same design fault which existed in the Herald of Free Enterprise and still exists in most European ferries in service today. However, the immediate or direct cause of the accident, although due to entry of water through the bow doors, was entirely different. In this case, the design of the bow doors had not allowed for typical wave heights in the waters of the Baltic where the ferry was being operated. Damage to bow doors on similar designs of ferry in the Baltic had in earlier years led to a number of near catastrophes prior to this accident. There was awareness within the industry that such an accident could happen, yet in spite of this, the warning signs were ignored. A latent error in the ferry design coupled with particularly severe sea conditions on the night of the sailing, resulted in the sinking of the 'Estonia' with the loss of 852 lives.

The final case study in this chapter concerns a village outing to a water treatment plant which turned to tragedy on a summer's evening in Lancashire. In this accident, as with many others, it took a number of coincident causes coming together in a most unfortunate way, to result in the deaths of 13 members of the public and three workers due to a methane explosion in an underground chamber. The accident was investigated by the Health and Safety Executive (HSE) who concluded that the explosion had been caused when a 12 kilometre long water pipeline connecting two rivers had been left empty for a period of weeks, allowing a build-up of methane gas in a tunnel section. Not only had the designers and operators been unaware of the potential hazard, but also the design of the water treatment plant did not cater for a possible build-up of flammable gas. The company who designed the plant were ultimately held responsible for the tragedy and forced to pay substantial damages to the survivors and families of the disaster, causing the company to cease trading.

All the design errors discussed in this chapter took place a number of years before the accident that they caused. They are perfect examples of accidents waiting to happen. Each case study concludes with a human error analysis showing in every case that there was a lack of foresight by the designer. In all cases the design failed to take full account of reasonably foreseeable operating conditions with which the design should have coped. Information about the detailed circumstances of design errors that took place many years ago are rarely available. Design errors are often associated with knowledge based tasks and it is open to speculation as to exactly what the designer was thinking at the time the error was made. In any case, designs are rarely the product of a single person working in isolation, they are usually drawn up in the context of a larger project and will be subject at various times to review and scrutiny by other designers and by the client. The aim here is not to direct blame towards a particular designer, but to show how these dangerous latent errors can remain undiscovered and slip through the net over a period of many years until they are revealed by an accident.

8.2 The fire and explosion at BP Grangemouth

8.2.1 Introduction

BP's Grangemouth refinery and chemical plant in Stirlingshire, Scotland was subject to three separate accidents involving fires and explosions between 13 March and 11 June 1987. These accidents cost the lives of four workers and as a result, the company was fined a total of £750,000 for breaches of the Health and Safety at Work Act (HASAWA). One of the accidents, which took place on 22 March, is described below and is used as an example of a major accident, the root cause of which was a serious design error. It is also interesting because the subsequent investigation and inquiry into the accident failed to highlight the significance of the design error. Rather it concentrated upon the errors made by the operators, another example of wrongly attributed blame. The accident took place in the Hydrocracker Unit of the refinery and involved a release of hydrogen gas from a high-pressure (HP) separator vessel into a downstream low-pressure (LP) separator vessel which was not designed for the HP operation. The overpressurization of the downstream vessel resulted in an explosive rupture and the discharge of flammable gases to the atmosphere leading to a serious fire and death of a plant worker.

8.2.2 Description of the process

Figure 8.1 shows a simplified arrangement of the HP and LP separators. In normal operation, a hydrocarbon liquid collects in the bottom of the HP separator and, as it builds up, forms a layer. The liquid is discharged through a level control valve into the LP separator ensuring that the HP separator is maintained at a constant level.

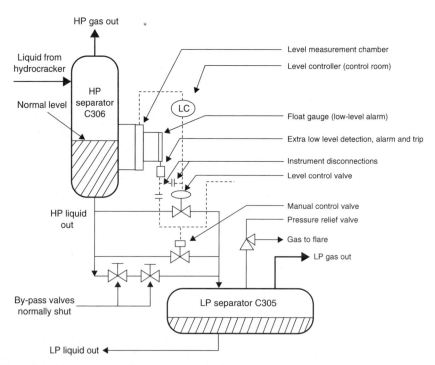

Figure 8.1 Arrangement of separators on hydrocracking process.

The level control valve for the HP separator not only restricts the flow of liquid from the vessel to control the level, but also acts as a pressure reduction valve for the liquid to pass from the HP to the LP section of the process. In addition, the layer of liquid in the HP separator acts as a seal between the two vessels to prevent HP hydrogen gas above the liquid passing into the LP system at a pressure for which it is not designed. In the event of a loss of level in the HP separator, a low liquid level detection device is provided to shut both the level control valve and a separate manual control valve, thus isolating the two separators. The LP separator vessel is further protected from overpressure by a pressure relief valve that discharges to a flare system in a safe location.

The designer had never intended the relief valve to be sized to accept the full flow of HP gas if there was a loss of liquid level in the HP separator. It had only been designed as a *thermal* relief to protect the vessel in the event of engulfment by fire. The volume of vapour generated by the worst-case fire was much less than the volume of gas that could pass from the HP to LP separator in the event of the level control valve being opened causing a loss of level. In fact, the regulations did not require the designer to provide more than a thermal relief valve. Yet, as discussed in Chapter 3, there is a reasonable expectation that a designer will take account of rules which are implicit within the situation itself. To do otherwise might be regarded as a knowledge based violation using the criteria set out in Section 3.3.4.3. In this case study, the failure by the designer is treated as an error rather than a violation, since the circumstances in which the design decision was made is unknown.

8.2.3 The accident

On 22 March 1987 the Hydrocracker plant was in the process of being started up. At about 02.00 hours, some 5 hours before the accident occurred, the plant had tripped as a result of a high temperature in one of the reactors which caused a delay to the start-up. It was decided to leave the plant on gas recirculation until the arrival of the day shift at 06.00 hours. At about 06.45 hours most of the operators were having breakfast in the mess room inside the control room building. At 07.00 hours the plant was subject to a violent explosion followed by an intense fire. The single casualty was a contractor who was killed on his way from the mess room to the plant. Most of the plant operators were uninjured because of their temporary absence from the plant at that time. The fires that raged for most of the day were brought under control by late evening. By then it become apparent that the LP separator had disappeared from its plinth. An investigation revealed that the steel pressure vessel had completely disintegrated with fragments being projected over a large distance. One 3-tonne fragment was found on the nearby foreshore about 1 kilometre away from the plant.

8.2.4 The investigation

Whereas the HP separator normally operated at about 155 bar, the normal operating pressure of the LP separator was only 9 bar. An inquiry into the accident by the HSE

(HSE 1989) investigated a number of mechanisms whereby the LP separator might have suffered such a catastrophic failure. It was calculated that the bursting pressure for the LP separator was about 50 bar and after the elimination of other possible causes it was concluded that it had been destroyed by an internal overpressure. Evidence strongly suggested that the accident had been caused by a breakthrough of HP gas from the HP separator due to the liquid level having fallen below the bottom outlet connection to the LP separator. The possible mechanisms by which the loss of level might have occurred were then examined.

A control valve receiving signals from the level controller (situated in the control room) normally regulated the level in the HP separator vessel automatically. However during the start-up this level control valve would be manually operated and left closed except when it was necessary to drain excess liquid from the HP separator into the LP separator. At such a time level the control valve would be opened momentarily while the level was observed to fall in the control room. When the desired level was reached, the valve would then be closed. A float gauge attached to the HP separator was used to measure the level and this sent a signal to the level controller. In addition, there was an extra low-level float chamber which sounded an alarm if the level in the separator fell below about 10 per cent of the level shown in the float gauge.

Originally the electrical signal from this low-level alarm was connected to two 'dump' solenoid valves in the pneumatic air supplies operating the level control valve and the manual control valve. Thus, if the level in the HP separator fell too low, then both valves would close and could not be opened until a normal liquid level had been re-established. However, some years prior to the accident, these solenoid valves had been disconnected and were no longer operative. As a result, when the level in the HP separator was being manually controlled, as at plant start-up, the protection of the LP separator from overpressure due to a gas breakthrough depended entirely upon the vigilance of the operator.

8.2.5 Report of the inquiry

The report of the investigations into the accident by the HSE draws a number of conclusions. The report is highly critical of BP's management of the process for allowing the system to operate without a high integrity automatic safety system to protect against gas breakthrough. Furthermore, there appeared to be little or no awareness of the potentially serious consequences to the plant if such a breakthrough occurred. While the direct cause of the accident is clearly human error by the operator in controlling the level in the HP system, it is equally clear that the systems supporting the operator were inadequate. This was an accident waiting to happen since a number of gas breakthrough incidents had occurred previously without serious consequences ensuing. However, it appears that these precursors either went unheeded or there was no appreciation of the serious potential consequences. The main vulnerability of the system was that the protection against gas breakthrough was dependent entirely upon the operator maintaining a liquid seal in the HP separator at all times during the start-up. This might have been possible, but on the morning of the accident he was seriously misled by a faulty level indication. As he opened the control valve in the control room to lower the level in the HP separator, there was no apparent fall in level. This caused the operator to open the control valve more and more expecting the level in the separator to fall. In fact unseen by the operator the level was actually falling. It continued to fall until a gas breakthrough occurred. It was an entirely predictable systemic human error that would certainly occur at some time.

There should have been a number of levels of defence against such an inevitable human error. The first level of defence should have been the low-level alarm operating independently of the level indication used by the operator to control the level. This would have warned the operator that the level had fallen too low. In the event, although this alarm apparently sounded, it was cancelled and disregarded since it had in the past proved unreliable due to a build-up of wax deposits. The next level of defence was the connection of the extra low-level switch to the solenoids on the two level control valves. This should have caused them to close or remain closed in the event of a low level. This would have prevented the operator draining the separator completely. However, these systems had been disconnected for some considerable time before the accident and although the operator was aware of this, it did not prevent him from making the error. While the operators might have been aware that gas breakthrough could occur, they may not have understood the potential consequences for the downstream equipment. However, even if there had been knowledge of the risk, it is not sufficient to rely on this to ensure that an operator never makes an error.

8.2.6 The design error

The capacity of the pressure relief valve on the LP separator should have been the final level of defence. This valve should have been sized to pass the maximum flow of gas from the HP separator with one or all of the level control/trip valves open. Since this device is a spring-loaded mechanical valve, it cannot be over-ridden as easily as the low-level trip system. One of the possible reasons why this ultimate defence was not deployed in the design was that the size of the pressure relief lines would have had to be increased considerably to cope with the extra flow of gas. This was a cost and convenience consideration that may have played a part in the design decision. Although the discharge of a large amount of HP gas to the flare might have been undesirable, it was the ultimate protection against failure of the safety system based on the closure of the control valves.

This design omission is the root cause of the accident. If adequate protection had been provided then the gas breakthrough might still have occurred, but would not have resulted in catastrophic failure of the LP separator. The only consequence would have been a harmless discharge to the flare system where the gas would have been safely burned off. Although the HSE report into the accident refers to this omission, it concludes that the operator failures that led to the gas breakthrough were the principal cause. However, rather than place the blame for the accident upon the operators, the HSE report holds the company to account for the failure, mainly because it had allowed the installed safety system to be bypassed. The design error was given little prominence in the report and the company that designed the faulty system was not criticized. The most important lesson to be learned is that a designer must always take account of probable human errors and the possibility that his best intentions may at some stage be defeated by human actions.

8.2.7 Conclusion

8.2.7.1 *General*

It must be concluded that the root cause of this accident was not the operator error that led to a loss of the liquid seal in the HP separator, but a serious failure of the plant designers to cater for a situation that was certainly bound to happen during the lifetime of the plant. It indicates the need for designers to understand the vulnerability of their designs to simple human error. It might be claimed that no plant can be completely foolproof, but in this situation all that was necessary was a little foresight in examining the possible scenarios that could lead to vessel failure due to a gas breakthrough. At the same time the reliability of the systems provided to prevent gas breakthrough and its consequences should have been assessed. The possibility that these systems might not work or would be bypassed should have been considered and the consequences assessed. In the event, the designer provided thermal relief on the LP separator as he was required to do, but this was inadequate to cope with a major gas breakthrough. It would appear that the fault was not picked up by any hazard analyses that might have taken place. It is to some degree the joint responsibility of the designer and owner/operator of the process to collaborate on this analysis, on the principle that

Table 8.1 Human error analysis of the Grangemouth explosion

Cause	Description	Type of error or other failure	Systemic cause or preventive action
Direct cause	Operator allowed level to fall in HP separator until gas breakthrough occurred	Active error – continued to open valve even when level indicator seen not to be responding	Faulty level indication and absence of error recovery (i.e. low-level trip was bypassed). Operator training in potential hazards.
Root cause	LP separator relief valve was not designed for maximum flow of gas from HP separator with the level control valves full open	Latent error at design stage	Hazard and operability studies (HAZOPS) on design to identify operating hazards (due to human error) and appropriate measures to protect against risk
Contributory causes	HP separator low-level trip system was permanently bypassed	Violation of good chemical plant practice by maintenance department	Independent audits of availability of safety protection systems
	Lack of operator awareness of danger of gas breakthrough	Active error – as for direct cause	Operator training

merely following the prescribed rules may not be sufficient to cater for all contingencies. Such an analysis should also include identification of critical human errors and the measures required to either prevent the errors or to limit their consequence. In this case, the provision of a properly sized pressure relief would not have prevented the error but would have limited the consequences and prevented the accident.

8.2.7.2 *Human error analysis*

Table 8.1 summarizes the main causes of the explosion and types of human error involved, together with the systemic cause of the human error. The latter indicates possible corrective or preventative actions, which might have been carried out to reduce the likelihood of the error or other failure.

8.3 The sinking of the ferry 'Estonia'

8.3.1 Introduction

The loss of the Estonian registered RORO passenger ferry 'Estonia' is a little remembered event in the annals of recent marine accidents, yet it resulted in five times the loss of life of the Herald of Free Enterprise disaster described in Chapter 7. Many of the features of the disaster were common to the Herald of Free Enterprise. In particular, the vulnerability of European RORO ferries to the entry of water onto the car deck and the resulting instability of the vessel. In a similar fashion to the Herald of Free Enterprise, the ferry capsized very quickly after the car deck had been flooded, not even allowing time for passengers to be mustered at the lifeboat stations for an orderly evacuation. The main difference between this and the Herald of Free Enterprise disaster was that the 'Estonia' was ploughing through heavy seas when the bow door mechanism failed. Most of the passengers who died were trapped below deck in their cabins by the rising water. The freezing water killed many of those who were able to escape to the upper decks because of insufficient time to launch the lifeboats. The other main difference to the Herald of Free Enterprise disaster is that the failure was not caused so much by an active human error, but by a design error in the attachment mechanism for the bow doors. However, in common with most of the case studies in this book, the accident was not caused by a single error occurring in isolation, but by an unfortunate combination of circumstances and errors which came together to cause in this case, Europe's worst ever ferry disaster.

8.3.2 The accident

The 'Estonia' departed at 19.15 hours on 27 September 1994 on a scheduled voyage across the Baltic Sea from Tallin, in Estonia to Stockholm, in Sweden. On board were 989 people, of whom 803 were passengers. Moderate sea conditions prevailed initially but the wave size increased once the ship had left the more sheltered coastal waters. The ferry was fully laden that night and had a slight list to port due to an uneven cargo disposition made worse by a strong wind that bore on to the starboard side. It is reported that an increase in the roll and pitch of the vessel caused some sea-sickness among passengers. By about 00.25 hours the 'Estonia' had established a course of 287 degrees, the speed was about 14 knots and the vessel was encountering heavy seas on the port bow. Shortly before 01.00 hours, a seaman carrying out his scheduled inspection of the car deck reported hearing a metallic bang coming from the area of the bow doors as the vessel ran into a heavy wave. This was reported to the sec-ond officer on the watch and another seaman was ordered to investigate. He did so by going to the loading ramp to check the indicator lamps for the visor and ramp locking devices. Everything appeared to be normal. However, over the next 10 minutes unusual noises were also reported by passengers and by off-duty crewmembers. The noises were sufficiently loud to be heard by passengers in their cabins.

Again the seaman was dispatched from the bridge to the car deck to investigate the noises. On this occasion he never reached the car deck. At about 01.15 hours the visor separated from the bow, allowing the loading ramp to fall open. Hundreds of tonnes of water entered the car deck and the ship very rapidly took on a heavy list to starboard. The ship was turned to port and the forward speed reduced. By now all four main engines had stopped and the ship was drifting in heavy seas. As the list increased, the accommodation decks became flooded and by 01.30 hours the starboard side of the ship was submerged and the list was more than 90 degrees.

Within a few minutes of the start of the disaster the list had increased so much that passengers became aware that something serious had taken place. Most passengers attempted to leave their cabins in the lower part of the ship to ascend to the deck but many of them were trapped due to the flooding and were not able to evacuate in time. On the boat deck, those passengers who had left their cabins were issued with lifejackets. A number of passengers jumped or were washed into the sea as a result of the speed with which the tragedy was now developing. A very small num-ber of passengers were able to clamber into emergency life rafts that had floated clear of the ship. It was impossible to launch any of the lifeboats, partly due to lack of time and partly as a result of the severe list. Other passengers and crew were thrown into the freezing water. The ferry disappeared from the radar screens of ships in the area at about 01.50 hours, sinking rapidly, stern first. Before the ship foundered, two mayday calls had been sent out and about 1 hour after the sinking, four pas-senger ferries which were in the vicinity arrived at the scene together with rescue helicopters.

Helicopters and ships assisted at the scene and managed to rescue 138 people dur-ing the night and in the early morning. In the following days 92 bodies were recovered but most of the persons missing went down with the wreck which lay on the seabed at a water depth of about 300 metres. Later inspection of the wreck revealed that the bow visor was missing and the ramp was partly open. The visor section was later located on the seabed about a quarter of a mile from the main wreck. The final death toll in the disaster was 852 persons, making it the worst ferry disaster in modern European history, and the worst disaster to strike the new nation of Estonia for many hundreds of years.

8.3.3 The investigation

An investigation into the causes of the disaster was conducted by a Joint Accident Investigation Commission from Estonia, Finland and Sweden and its report published in 1997 (Meister A, 1997). The findings were that the vessel had suffered a capsize as a result of a major failure of the bow visor under the impact of heavy wave action. It was not considered that these waves were excessive for the waters in which the ferry was sailing. There was no apparent reason why the wave action should have caused failure of the bow visor allowing water to enter the car deck. The investigation commissioned a number of tests to measure the strength of the bow visor, which had been recovered from the sea, its hinges and locking mechanisms. It was discovered that the points where the visor, its locking devices, the deck hinges and the lifting cylinder mountings were attached had been subjected to a high tension due to wave impact on the underside of the visor. This pounding action had caused the attachments to break due to their inadequate design strength, allowing the visor to become completely detached. This explained why the visor was found on the seabed some distance away from the main wreck.

Under low wave conditions it was found that the design strength of the attachments were perfectly adequate. However, due to the overhang, or rake, of the visor, larger waves impacting upon the visor would exert a powerful turning moment placing the hinges and other mechanisms under excessive strain. The first loud banging noise heard by the crew and passengers was thought to have coincided with the partial initial failure of the lower visor locking mechanism. The later noises and metallic bangs were probably heard after the remaining locking devices had failed completely allowing the visor to fall open. It then began to impact intermittently with the hull of the ship due to wave action causing it to move backwards and forwards. Finally the visor broke free to tumble forwards into the sea. The time between the visor separating from the vessel until the start of water ingress onto the car deck was probably a few minutes. The total time from initiation of damage until sinking may only have been about 10–20 minutes.

8.3.4 The causes

The direct cause of the loss of the 'Estonia' was a capsize due to large quantities of water entering the car deck. This caused a major loss of stability and subsequent flooding of the accommodation decks. Contributory to this was the full-width open vehicle deck allowing water to accumulate and flood to one side contributing to the rapid listing (see Herald of Free Enterprise Accident in Chapter 7 for explanation and similarities). The disaster was made worse because prior to the engines being stopped the vessel had turned to port, exposing the open ramp to the sea. This also exposed the side of the vessel which was listing to wave impact so that the windows and doors were breached leading to further progressive flooding.

The root cause of the disaster (according to the report of the Investigating Commission) was the under-design of the bow visor and ramp attachment mechanisms. This had the most severe consequence for the progression of the accident. In this respect the vessel did not comply with the Safety of Life at Sea (SOLAS) regulations (again see Chapter 7) concerning the design of the bulkhead, although these designs had been accepted by the national inspecting authority before the ship was commissioned. One of the problems was that the design, manufacturing and approval process did not appear to consider the visor and its attachments as critical items regarding ship safety. This was not the case in other parts of the world where bow visors and their attachments were more rigorously designed and strongly constructed.

8.3.5 The aftermath

After the report into the sinking of the 'Estonia' was published, a great deal of controversy took place concerning the validity of its findings. The German designers of the bow visor and ramp mechanism, believed that the failure of these components was not the root cause of the accident, but that the ferry had been holed on the starboard side prior to the accident allowing water to enter the vessel. No reason was advanced as to how this might have happened except that the ferry was operating in near storm conditions. The designers suggested that the visor had become detached as a result of the accident and was not therefore its original cause. Furthermore, the designers believed that the visor and ramp mechanisms had not been properly maintained by the vessel owners, and as a result had become seriously weakened contributing to the loss of the visor. This was supported by photographs taken by a passenger on the 'Estonia' a few weeks before the accident clearly showing deficiencies in the hinge mechanism. It was also discovered that the crew had experienced numerous problems with the closure of the visor, and that these had been reported but not corrected. After the inquiry report was published it was revealed that the ferry had never been designed for service outside coastal waters, and should not have been allowed to operate in the sea conditions that prevailed on the night of the disaster. This may also have contributed to the sinking.

In the immediate aftermath of the disaster, the governments of the three countries expressed an intention that the ferry would be raised from the seabed since it was lying in shallow water. However, this decision was later revoked and the wreck was designated a national grave. To prevent the possibility of interference thousands of tonnes of pebbles were deposited on the site with a plan to eventually construct a concrete sarcophagus to contain the wreck. Unfortunately, this will also inhibit further investigation of the wreck to decide once and for all whether in fact the vessel had been holed prior to the visor becoming detached.

There are similarities here with the Flixborough disaster described in Chapter 7. Although the investigation was able to identify and report upon the causes of the accident, the real cause, even to this day, remains in doubt and the final story may never be known. Whatever the cause, the findings of the inquiry did not place the government or approval authorities in a particularly good light. There had been numerous bow visor incidents prior to the 'Estonia' accident on vessels built both before and afterwards. One of these had involved the 'Diana II', a near-sister vessel to the 'Estonia', only a few years before. In all these incidents, bow visor and ramp problems had allowed water ingress on to the car deck. In all cases, the ferries had managed to reduce speed and return to port averting a similar tragedy. These incidents did not prompt any investigation or demands for strengthening of the bow door and visor attachments on existing vessels. The reason for this was that there was no system for collecting data about incidents and collating and disseminating these through the industry so that lessons could be learned.

8.3.6 Conclusions

8.3.6.1 *The design of the bow door mechanisms*

Irrespective of the reason for the bow visor and ramp attachment failure there are lessons to be learned regarding the interaction between designers and operators. Pressure is often placed upon a supplier to produce a low-cost design with incentives to get the completed project out of the door as quickly as possible. This can lead to a temptation to take shortcuts, often at the expense of future maintenance, operability and reliability of the system. Once the system has been commissioned, then the company that designed and built it will want to submit their invoices, be paid and walk away from the job as soon as possible. It is probable that the next system to be built will follow the pattern of earlier designs in order to save time and cost. If there has been no feedback from the previous design concerning operational problems arising, then the same problems will arise again. Experience of previous failures of bow visor and ramp mechanisms had not been disseminated through the industry, nor reported back to the designers. It was therefore only a matter of time before a recurrent problem turned into a disaster.

Information is not available about whether the visor mechanisms were adequately maintained. Even if they were, it is not known whether the attachments would still have failed because they were grossly under-designed. The controversy following the investigation does however illustrate the difficulties of an inquiry arriving at an objective decision when the parties who conduct it have a vested interest in the outcome. There is no evidence that the governments who set up the commission had any influence over its findings, or that those findings were biased in order to deflect blame away from the government and its approval authorities. However, when the government decided not to lift the ferry from the seabed, then there was inevitable speculation that they were afraid of what might be found. It had the effect of prolonging the controversy. In this respect there are similarities to the Flixborough investigation and inquiry (see case study in Chapter 7) which also neglected to follow up lines of investigation, arousing suspicions that there were certain aspects of the design which were not open to scrutiny. Nevertheless, the Commission recommended that all attachment points for a bow door should be regarded as highly loaded design items and subjected to detailed strength analysis and future improvement may be expected.

8.3.6.2 *The need for hazard analyses*

A wider lesson to be learned from this case study is that where operating experience is limited and very few design rules exist, it is important to carry out hazard and consequence analyses of failures of critical components. This is certainly the case in the aircraft and nuclear industries, and unless such studies are carried out in these industries then the systems cannot be licensed for use. It would seem that in this case, even a very simple analysis of the consequences of failure of the bow visor mechanisms, would have shown that the design was critical to safety. The consequences of even small quantities of water entering the car deck were well known after the much-publicized Herald of Free Enterprise disaster some 9 years earlier. As with the earlier disaster, proposals for retroactive upgrading of existing ferries were met with resistance from the industry namely, the International Maritime Organization (IMO), and maritime classification and approval bodies.

The overriding impression is that, 9 years after the Herald of Free Enterprise disaster, very little had changed in the area of marine safety, particularly regarding lack of retrospective legislation about the design of RORO ferries. As a case study for human error in design, the main lesson to be learned is that designers need to look beyond the drawing board to the uses and abuses to which their designs may be subjected in the long-term. With hindsight, it seems inconceivable that a designer would not allow for the effects of repeated wave action upon the overhanging section of the visor. It seems that the visor attachments were designed for calm conditions, but in heavy seas there would be overstressing of the components locking and holding the visor in position. Although it might be possible to specify that a ferry must always sail in calm coastal waters, it is possible that one day it will face storm conditions. To place such a restriction upon the owner would seem unacceptable, in much the same way that the integrity of the LP separator at Grangemouth came to depend entirely upon the operator not making an error.

8.3.6.3 *The role of officers and crew*

The investigating Commission were somewhat critical of the officers on the bridge because they were not aware that the list developing on the ship was due to water entering through the open bow door. They were also criticized because they did not make a speed reduction prior to investigating the reports of noises from the bow area. The example of the Herald of Free Enterprise again comes to mind. The officers on the 'Estonia' were unable to see the bow visor from the bridge, although there was a TV monitor which would have shown water entering the car deck had they observed it. In mitigation, the indicating lamps which showed whether the visor was locked, were still showing it locked even after the visor had tumbled into the sea, due to an unreliable design. As the visor failure was so rapid and severe, it seems difficult to attach any significant blame to the master or crew, since they had hardly any time to investigate the unusual noises from the bow before the ferry was fatally compromised. The principal difference between previous ferry incidents and the incident which resulted in the sinking of the 'Estonia', was the vagaries of the weather and fact that the ferry did not slow down or head for calmer waters.

8.3.6.4 *Human error analysis*

Table 8.2 summarizes the main causes of the disaster and types of human error involved, together with the systemic cause of the human error. The latter indicates possible

Table 8.2 Human error analysis of the 'Estonia' sinking

Cause	Description	Type of error or other failure	Systemic cause or preventive action
Direct cause	Bow visor became detached from its fastenings in severe weather causing water to enter car deck	Bow visor attachments not strong enough to resist upward forces due to wave action	See root cause
Root cause	Design of bow visor was inadequate for sea conditions in waters where ferry was operating	Latent error at the design stage – failure to design for worst conditions or	System to ensure design specification based on operating conditions
		Management error – owners operated ferry beyond its known design limitations	System to ensure operating conditions never exceed design specification and inspection of safety critical features
Contributory causes	Helmsman/officer did not reduce speed after receiving reports of noises from bow doors	Active error	Lack of information about previous incidents which could have warned the helmsman to take action
	Inherent instability of RORO ferries with water on car deck	Design (latent) error	Industry had failed to take cognisance of the proven instability of this type of ferry with only a few inches of water entering the car deck. Refusal to consider retrospective implementation of improved standards for freeboard.
	Bow visor attachments in poor condition	Maintenance error	Improved maintenance of safety critical components

corrective or preventative actions, which might have been carried out to reduce the likelihood of the error or other failure.

8.4 The Abbeystead explosion

8.4.1 Introduction

The Abbeystead explosion occurred on 23 May 1984 at the Abbeystead water pumping station of the North West Water Authority (NWWA) in Lancashire. It resulted in the deaths of 13 members of the public and three employees of the NWWA. Most of the deaths resulted from severe burns incurred as a result of the accumulation and subsequent ignition of methane gas which leaked into a valve house in the pumping station during a guided tour of the installation. The valve house at Abbeystead was a concrete structure set into a hillside at the outfall end of a pipeline connecting the rivers Lune and Wyre. The purpose of the pipeline was for water transfer between the two rivers to provide additional supplies for the industrial towns of north-west

Lancashire. In normal circumstances the water authority might have provided a reservoir to supply the additional water needs of the area, however, permission for this was difficult to obtain in this area of outstanding natural beauty. The pipeline was some 12 kilometres long but for 7 kilometres of its length had to pass through a tunnel bored under Lee Fell, a 137 metre peak on the edge of the forest of Bowland. This 2.6 metre diameter tunnel was lined for most of its length with porous concrete, a design which allowed water under pressure in the surrounding rock strata to seep into the tunnel. The system was designed to allow up to 280 million litres of water per day to be pumped augmented by water ingress from the rocks of 12 litres/second. At the receiving end, the water flowed into the valve house at Abbeystead and into an underground chamber from where it passed over a weir into the River Wyre. The Abbeystead water pumping scheme was of a novel design and was felt sufficiently important at that time to be officially opened by Her Majesty the Queen in 1980.

During the first few years of operation, the pumping station had low utilization and transferred no more than about 90 million litres a day, operating every other day on average. In spite of this, villages along the lower Wyre Valley, downstream of the Abeystead outfall had experienced unusually severe winter flooding since the station had opened. The NWWA had strongly denied that the flooding was due to the use of the water pipeline. Faced with adverse local reaction arising from the flooding, the NWWA decided to arrange a public relations exercise to restore confidence. An event was arranged whereby a small group of villagers and members of the parish council of St Michael's-on-Wyre (a village some 24 kilometres downstream from the pumping station) would visit the Abbeystead valve house. The purpose was to demonstrate to them that the flow of water passing through the station could not possibly cause the River Wyre to flood to the extent that had occurred.

The visit was arranged for Wednesday 23 May 1984, and a total of thirty-six villagers arrived on a bright summer evening to be greeted at Abbeystead by eight officials of NWWA. The scene was set for a unique and unpredictable disaster that was to bring tragedy to the lives of many in that small village. It was unusual and unfortunate that in May of that year there had been a severe drought. It was unusual because Abbeystead is located in the north-west of England, an area of high rainfall; it was unfortunate because, as a result of the drought, the pumps at the River Lune had not been operated during the previous 17 days. This was to have a catastrophic effect when they were eventually started.

8.4.2 The accident

The Abbeystead pumping station had been purposely built into the side of a hill and the roof covered with turf in order to blend into the countryside. Concrete beams supported the turf roof. On arrival, the visitors crowded into the first of the two main rooms of the Abbeystead pumping station. The so-called 'dry room', which housed the various valve controls, also included a small office and toilet. This room opened through double doors into an even smaller 'wet room'. In here, the floor consisted of a metal grating below which could be seen the main water discharge weir and distribution chambers.

The cramped surroundings of these rooms were never intended to accommodate more than a few people at a time. After a short talk by a NWWA manager a request was sent by telephone to start the water pumps at the other end of the pipeline. The party waited to observe the flow of water over the weir in the discharge chamber below the grating. This weir regulated the quantity of water passing into the River Wyre. The demonstration was to be followed by a short walk from the pumping station to the discharge point into the river to observe the outflow and compare it with the flow of the river. As the pipeline was empty, the visitors were subjected to a long wait. During this time, some of the visitors became bored and wandered outside the building.

Just before the flow of water arrived at the chamber, a blue glow was observed coming from below the metal grating and the visitors felt an intense build-up of heat inside the room. This was followed a few moments later by a huge explosion that threw the 11-inch square concrete beams, forming the roof of the chamber, high into the air. Some of the visitors inside the room were thrown into the water, others were crushed by collapsing debris and most received severe skin burns. The emergency services were alerted by local people who had heard the sound of the explosion and soon arrived at the scene. The injured were conveyed by ambulance to the closest hospital in nearby Lancaster. There was not a single person present at the pumping station who was not injured in some way. Eight immediate fatalities resulted from blast and crush injuries, and eight people died later in hospital, mainly as a result of burns.

It was later established that the explosion had been caused by the ignition of a mixture of methane and air due to an accumulation of gas in the pipeline which had been displaced and pushed forward into the valve house by the flow of water. It was not immediately obvious how such a methane accumulation in the pipeline had occurred. No provision had been made by the designers of the pumping station to cope with a build-up of inflammable gas in the system because none was expected. The fact that the tunnel had been empty for a period of 17 days before the explosion occurred was an indication that an unknown source of methane existed within the tunnel system. Neither was any obvious source of ignition identified. No faults were found in electrical equipment in the building. It is easily possible that one of the visitors may have tried to light a cigarette, or that somebody's footwear in contact with the metal grating had created a spark. The NWWA were so confident that a flammable atmosphere could not arise, that there was not even a specific prohibition against smoking in the valve house.

8.4.3 The investigation

Almost immediately an investigation into the disaster was opened by the NWAA. The following day inspectors from the HSE arrived to undertake their own independent inquiries. Tests were carried out in the tunnel and on the valve house atmosphere and it very quickly became obvious that methane was present in significant quantities. The source of the methane was at that stage unknown. As more information became available it was obvious that the designers and builders of the installation had not been aware of the true geological nature of the strata through which the tunnelling had taken place. The tunnel wall had been deliberately constructed of porous concrete since there seemed little point in trying to exclude the ingress of high-pressure water from the surrounding rock because this water would augment the supply being pumped between the rivers.

It was also intended that the tunnel would remain full of water when it was not in use and an instruction to this effect was incorporated into the operating manual for the pumping station. However, in the early period of operation it had been discovered that if the pipeline remained full, then an accumulation of sludge would cause discoloration of the water in the River Wyre when pumping was started. Thus, in violation of the operating instructions, and unknown to NWWA management, a valve at the end of the tunnel used to stop the flow of water was always left slightly open so that the tunnel would drain into the river leaving it permanently empty.

It was at first suspected that the methane may have arisen from the decomposition of rotting vegetation in the tunnel. However, it did not seem possible that this source could have released such a large quantity of methane over the 17 days that the tunnel was empty. In fact, carbon isotope tests soon revealed that the methane was in fact of the order of 30,000 years old, which meant that it must have a geological origin. It quickly became clear that the water leaking into the tunnel from the surrounding rock was rich in dissolved methane. While methane is not normally soluble in water at atmospheric pressure, under the intense pressure within the rocks, the solubility is sufficiently high for the water to contain a high concentration of dissolved methane. However, as soon as the water enters the tunnel, the reduction in pressure causes the dissolved methane to be released into the atmosphere as a gas. During the period that the tunnel was empty, it became a sealed chamber into which methane gas could accumulate. When the water pumps at the River Lune were started on the evening of the accident, this methane atmosphere was forced along the tunnel to emerge at the Abbeystead valve house where it ignited.

Although the contractors building the tunnel had sunk a number of boreholes to test geological conditions along the route, these never produced any evidence of methane. However, the HSE investigation revealed that the equipment the builders had used to carry out tests for inflammable gases during tunnel construction was incapable of differentiating between methane and other gases.

The HSE investigation report (Health and Safety Executive, 1985) concluded that the root cause of the disaster was that the possibility of a methane rich environment had been overlooked by the designers. They had not considered whether it was possible for methane in significant quantities to be dissolved in water that leaked into the system. It is probable that the explosion would not have occurred had a proper venting system been designed and constructed at the valve house. Although a venting chamber had been provided at the lower end of the pipeline this chamber had inexplicably been vented back into the 'wet room'. The effect was that there was no atmospheric relief to safely discharge any gases which might accumulate in the system.

In spite of the findings of the HSE investigation, the parties involved in the design, construction and operation of the facility continued to insist that that there had been no evidence for the presence of significant quantities of methane, and therefore they were not negligent. Claims for compensation to the victims were disputed on this basis. Over and above the omission of the designers and operators to allow for the presence of methane gas building up in the system, the HSE investigation also highlighted a number of other failings in the way the system was operated. The plant had been operated in violation of the instructions in the operating manual provided by the designers without any consideration of their impact upon the safety of the system. In addition, the operators had not been made aware of the significance of certain features of the pumping installation, training was not considered to be adequate and competence testing had not been carried out.

8.4.4 The aftermath

The HSE investigation report concluded that both the designers and operators 'believed and still believe that methane was not emergent from the strata in quantities which were significant'. On the basis of this both parties denied liability for injury suffered by the visitors to the plant. Hence the survivors of the disaster decided to undertake their own investigations before taking out a civil action against the designers, builders and operators of the Abbeystead pumping station. To further this action, the survivors hired an expert geological engineer, whose inquiries immediately revealed that coal seams had been worked in the area some 200–300 years before. He found that the whole area had been subject to coal mining during the 17th and 18th centuries, and Coal Board records confirmed this.

The presence of coal is often associated with varying amounts of 'fire-damp', or methane gas, which of course would have presented a major explosion hazard at that time. The methane migrates through the porous rock structures until it reaches an impervious layer where it is trapped and dissolves in water passing through the rock. If the designers and builders had carried out a more thorough investigation of the geology of the area, it is almost certain that this would have revealed the possible presence of methane. When the civil action for damages came before the High Court in 1987, this new evidence was presented. Also presented was evidence from a Polish ex-mining engineer who had worked on the construction of the tunnel. His testimony stated that on three occasions during the construction, he had made reports to his superiors that methane was leaking into the tunnel workings. On one occasion there had been a major breakthrough of water into the tunnel, the presumption being that this had arisen from a spring. However, the Polish engineer had pointed out that the water was laden with dissolved methane being released in the gaseous phase as the water entered the tunnel at reduced pressure.

In the face of this new evidence, the constructors and designers of the system were found to be negligent and thus liable to pay damages in the High Court. In the end, the designers alone were found liable, particularly in respect to their failure to provide adequate venting of methane and other dangerous gases which might accumulate in the valve house. This was in violation of design codes and best practice. The result was that the designers were ordered to pay £2.2 million in compensation for deaths and injuries to the bereaved and to the survivors. The company who designed the installation eventually went out of business.

8.4.5 Conclusion

8.4.5.1 *General*

This case study illustrates yet again how a latent error at the design stage is revealed by a catastrophic event some years after the system has been commissioned. The important omission of an adequate venting system in the design of the valve house might have been discovered if a more intensive geological survey of the surrounding rock strata of the tunnel had been carried out prior to construction. In fact even the investigation by the HSE did not reveal the true nature of the strata. This was revealed only when further investigations were commissioned by the survivors and bereaved in their pursuit of compensation for injury and death long after all other investigations had ceased. The later investigation revealed that evidence of methane in the water leaking into the tunnel was available to those constructing the facility, but had not been followed up.

Even in the absence of evidence to suggest the presence of methane, it is unusual that an underground facility could be designed, constructed and accepted by the client, without adequate venting arrangements. The failure of the official investigations to reveal the existence of coal mining in the area indicates yet again the importance of ensuring that the scope of an investigation is as broad as possible. As with the Flixborough inquiry, described in Chapter 7, the investigation should not exclude any events even if at first sight they seem to be unrelated to the accident. In this case, the evidence of a Polish ex-mining engineer, who had seen and reported evidence of methane gas dissolved in the water issuing from the strata, would have been crucial to discovering the cause of the disaster had it been uncovered by the investigating authorities. Had the information been acted upon earlier, it is possible the accident might have been prevented.

As with other design errors in this chapter, details do not exist of how the error actually came to be made by the designer and cannot therefore be part of this analysis. It would appear that the omission was not discovered by the design review and checking

Table 8.3 Qualitative human error analysis of the Abbeystead explosion

Cause	Description	Type of error or other failure	Systemic cause or preventive action
Direct cause	Accumulation of methane from tunnel and ignition in valve house	Violation of operating rules (tunnel was left empty of water)	Operator training
		Management error	Supervisory oversight and checking of work practices
Root cause	Design did not cater for possible accumulation of dangerous gases by provision of proper venting to atmosphere from the valve house	Latent error at the design stage	HAZOPS to identify all possible hazards and measures to protect against unacceptable risks
Contributory causes	Presence of public in an unsafe environment	Management error	Initiate company policy for admission of public to dangerous facilities

process which, it is hoped, the company undertaking the design would have conducted. Nor was the omission detected or questioned by the client on receipt of the design details. This emphasizes the importance of checking for latent errors in design and maintenance activities. This is because the checking process is usually the final opportunity to detect a latent error before it causes an accident.

8.4.5.2 *Human error analysis*
Table 8.3 summarizes the main causes of the disaster and types of human error involved, together with the systemic cause of the human error. The latter indicates possible corrective or preventative actions, which might have been carried out to reduce the likelihood of the error or other failure.

References

Health and Safety Executive (1985). *The Abbeystead Explosion. The Report of the Investigation by the Health and Safety Executive into the Explosion on 23 May 1984 at the Valve House of the Lune/Wyre Transfer Scheme at Abbeystead*, London: HMSO.

Health and Safety Executive (1989). *The Fires and Explosions at BP Oil (Grangemouth) Refinery Ltd*, Sudbury, Suffolk: HSE Books.

Meister, A. (1997). *Final Report on the Capsizing of the RO-RO Passenger Vessel MV 'Estonia'*, Tallin: Government of the Republic of Estonia.

9

Maintenance errors

9.1 Introduction

Most maintenance errors have delayed consequences and fall into the category of latent errors. They therefore have some similarity to management and organizational errors as described in Chapter 7. Maintenance errors occur in the workplace rather than at management level although it is often the case that the root cause lies within the management system that allow the error to happen. This is demonstrated in the case studies described below.

The first of the case studies is from the aviation industry and could well have resulted in the loss of an aircraft of the Royal Flight. Fortunately, the error was recovered and resulted in a dangerous incident rather than an accident. This is unusual, because serious maintenance errors involving aircraft rarely allow the opportunity for recovery from a failure of a critical system once the aircraft is in flight. For this reason, critical aircraft systems usually have multiple redundancy provisions such that if one system fails then there will usually be a backup system available to take over. An obvious example of this occurs with 2, 3 or 4-engine commercial aircraft, where the aircraft is designed to fly on one engine alone. Apart from engine redundancy, there are many redundant components built into the critical systems on an aircraft such as hydraulics, instrumentation, navigation and communications. However, as the case study in Section 9.2 below will illustrate, redundant systems are very susceptible to latent errors that exhibit dependency. It was due to latency coupled with dependency that a series of errors was able to fail all four engines of the British Aerospace (BAe) 146 aircraft of the Royal Flight.

The second case study concerns a high-profile train derailment at Hatfield near London that claimed four passenger lives in 2001. The derailment was one of six major rail accidents that have occurred since privatization. It brought into prominence a number of critical issues arising out of the way maintenance is conducted under privatization. The final report of the investigation into the accident had, at the time of writing, not been published and it is therefore not possible to draw detailed conclusions about the errors which actually occurred and which were the direct cause of the accident. However, based on the interim reports of Her Majesty's Railway Inspectorate (HMRI), it is possible to draw some conclusions about the root causes of the accident. The case study is an extension of the themes discussed in Chapter 7, concerning organizational and management errors associated with rail privatization. The case study shows how problems at a high-level in the organization may not manifest themselves until an accident occurs caused by errors at the operational level. The case studies conclude with a brief analysis of the more recent rail accident at Potters Bar, a few miles down the track from Hatfield.

9.2 Engine failure on the Royal Flight

9.2.1 Introduction

The BAe146 aircraft is a short to medium range jet airliner built in the UK and in its commercial configuration carries around 100 passengers. When the aircraft was first brought into service it was nicknamed the 'whisper jet' because of its low noise operation. Dut to this and the short take-off and landing characteristics, it is commonly used on commuter services operating from city centre airports such as London City which typically have short runways and are close to centres of population. The BAe146 is equipped with four Textron Lycoming turbofan engines; it has a maximum cruising speed of 423 knots and a range of about 1200 nautical miles. The aircraft is in widespread commercial use around the world. Four of these aircraft were adapted for the use of the British Royal Family because of their fuel economy and the perceived high reliability of a four-engine configuration. They are operated by an RAF squadron based at Northolt, west of London. The Prime Minister, members of the Cabinet and senior civil servants on official business also use the BAe146s of the Royal Flight.

The case study concerns an incident involving one of the aircraft of the Royal Flight when it was forced to make an emergency landing as a result of a failure of three of its four engines. It provides an excellent example of a maintenance failure resulting from a latent error exhibiting high dependency as described in Section 5.4. Fortunately, when the accident occurred the aircraft was on a training flight from Northolt. However, a few days earlier, the Duke of York had travelled as a passenger on this particular aircraft. Shortly after this flight the aircraft was submitted for some minor maintenance work, when a number of catastrophic human errors were made.

9.2.2 The accident

The BAe146 aircraft took off from Northolt for a training flight on 6 November 1997 flown by a trainee captain with a senior RAF instructor and a flight engineer on board. Fifteen minutes into the training flight, the three-man crew observed that the oil quantity gauges for Nos 2, 3 and 4 engines were indicating empty. The oil quantity gauge for the No. 1 engine was showing less than a quarter full. This was followed almost immediately by a low oil pressure warning light for the No. 3 engine which was shut down.

An emergency was declared and the pilot requested immediate clearance to land at the closest suitable airport, Stansted in Essex. Within a few minutes of the initial low pressure warning light, further warning lights then came up for the Nos 2 and 4 engines. The response of the crew was to shut down No. 2 engine followed immediately by the No. 4 engine. As the aircraft was coming in to land at Stansted using the power from the remaining No. 1 engine, this engine was also showing low oil pressure.

9.2.3 Cause of the accident

In November 1997, a private maintenance contractor carried out maintenance on all four engines of this Royal Flight aircraft. The maintenance was routine and involved taking samples of oil from the engines for spectrometric analysis, a procedure conducted every 50 flight-hours. The procedure also included changing the magnetic chip detection plugs (MCDPs) in all four engines. These plugs are designed to detect contamination of the oil with metallic particles, which could indicate potential mechanical problems with the engine. The plugs are installed with an 'O' ring oil seal between the plug and the engine casing to prevent leakage of oil while the engine is running. The MCDPs are normally issued to the maintenance staff in kit form and include all the parts required to replace the plugs including the 'O' ring oil seals.

This particular maintenance operation took place during the night shift and maintenance staff who undertook the work were depleted in number. The initial snag arose when the supervisor in charge of the work was unable to find a technician to change the MCDPs. He therefore decided to undertake the work himself, a task with which he was not familiar. He immediately discovered that no MCDP kits had been made available by the day shift so he decided to assemble a kit for himself using various components from the stores. While collecting the parts, he failed to include a set of 'O' ring oil seals. The relatively simple assembly task was then undertaken but the plugs were fitted without an 'O' ring between the plug and the engine casing.

In normal circumstances, this error might not have been so serious, since after completion of maintenance by a technician, the work would be checked by a supervisor and signed off before the aircraft was released for service. As the normal technician was not present to carry out the work, the supervisor decided to ask another technician to sign off the work as if he had carried it out. The supervisor then signed the work off as checked although no second person check had been made. In this way the absence of the 'O' ring oil seals went undetected. The error was therefore made more likely by a violation of working procedures.

9.2.4 Conclusion

9.2.4.1 *Error latency*

The omission of the 'O' ring seals from the MCDPs is clearly a latent error. The consequence of the error only became apparent at a later time, in this case when the aircraft was in flight. By this time there was no possibility for recovery. The reason for the delayed consequence was that a major loss of engine oil did not occur until the engines had been started and the oil system was pressurized. Even then, the error could only be detected when the engine had been running long enough for most of the oil to have drained away through the unsealed gap between the plug and engine casing. Once in flight, the only possible means of mitigating the consequences was for the engine to be shut down immediately to protect it from further damage, but endangering the aircraft. If the engine had been left running, then the pilot risked the even greater danger of an engine fire due to overheating of the moving parts. The fact that the same error was made on all four engines indicates that, in addition to being a latent error, this is also an example of error dependency. It was this dependency that caused the same error to destroy the redundancy of a four-engine aircraft as described below.

9.2.4.2 *Error dependency*

Error dependency is described in detail in Section 5.4 and occurs when an initial error increases the probability of the same or a similar error being made later. In this case, according to the root cause classifications described in Section 5.4.2, this error can be defined as an externally induced dependent error. This means that the root cause of the errors is an external mechanism or performance shaping factor (as opposed to a psychological mechanism within the person undertaking the task as is the case of internally induced dependent error). In this case, the external mechanism was the omission by the supervisor to include the 'O' ring seals in the self-assembly MCDP kit made up of *ad hoc* components drawn from the stores. Due to his unfamiliarity with the task, the error went unnoticed and the scene was set for a dependent failure. There would seem to be two levels of dependency at work in this situation:

- *Task 1 – Making up the MCDP kit.* This comprised the four separate sub-tasks of making up a kit for each of the four engines. On making up the first kit, the 'O' ring seal was omitted. Having made this omission, there was a much higher probability that the same error would be made while assembling the other three MCDP kits. The probability of the dependent errors, having made the first error, was probably something approaching unity, that is, a near certainty.
- *Task 2 – Replacing the MCDPs.* This comprised the four separate sub-tasks of replacing the MCDPs in each engine. Once an MCDP had been replaced in the first engine without an 'O' ring seal, it was again a near certainty that the same mistake would be made with the other three engines.

The dependent errors described above are often referred to as recursive errors because of their close similarity. They are recursive because both the task and the error are exactly the same in each case.

There is also a form of dependency at work between Tasks 1 and 2 above, since the failure to include the 'O' ring seal in the kit seems to increase the probability that a failure to fit an 'O' ring seal will occur. As there was a complete absence of any checking of the work, there was no realistic chance of recovery from the error. Furthermore, because the supervisor manipulated the quality control system to obtain a signature,

Table 9.1 Qualitative human error analysis of the accident to the Royal Flight

Cause	Description	Type of error or other failure	Systemic cause or preventive action
Direct cause	Loss of oil from all four engines due to failure to fit 'O' ring seals between the MCDPs and engine casings.	Latent error Dependent error	Independent checking of quality of work to ensure that recursive errors are not made.
Root cause	Failure to adhere to maintenance procedures: – the maintenance kit was self-assembled and as a result the 'O' rings were omitted – the checking procedure (checking and signing off by a second person) was omitted –the quality records were falsified thus precluding error recovery.	Violation	Lack of management and quality controls. Failure of management oversight. The procedure when completed should be subject to checking and signing off by a second person, not carrying out the work.
Contributory causes	The person carrying out the procedure, although a supervisor, was not experienced in the task, and therefore more prone to error and less able to discover his own errors.	Active error	There is a possible cultural problem in the organization which should be investigated, i.e. undue pressure resulting in employees violating procedures in order to complete work on time.
	The person carrying out the procedure may have been under economic or job pressure to complete the work on time and tempted to bypass normal procedures in order to do so.	Management error	Management responsibility to remove excessive work pressures and restore correct priorities.

this made it even more likely that the error would not be picked up prior to the aircraft being put into service.

9.2.4.3 *Human error analysis*

9.2.4.3.1 Qualitative analysis

The qualitative human error analysis in Table 9.1 identifies all the main errors and error types which caused or contributed to the accident. For this case study it supports the quantitative analysis described in Section 9.2.4.3.2.

The latent and dependent errors are examined in more detail in the quantitative analysis described below.

9.2.4.3.2 Quantitative analysis

Once airborne, the aircraft would be capable of flying with three out of the four engines shut down. The loss of all four engines at the same time would have an extremely low probability of failure if independent causes are assumed. Chapter 5 describes the method of calculating the overall probability of failure of multiple components or

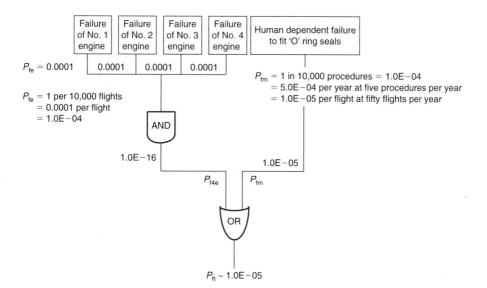

Figure 9.1 Fault tree for multiple engine failure on BAe146 of the Royal Flight.

errors assuming the separate failure probabilities are known and the events are independent. The method uses the multiplication rule to calculate the overall probability. The main problem with this is that if the events are not truly independent, then the overall probability calculated may be unrealistically low. In the case of human error probabilities, the problem was overcome by the use of human error limiting values (HELVs) as described in Section 5.3.3.3. A similar approach can be adopted here using the probability of a common cause affecting all four engines to offset the extremely small independent probability of four engines failing simultaneously during a flight. The fault tree for this scenario is shown in Figure 9.1.

It has been necessary to make a number of assumptions in the preparation of this fault tree, since information about the operation of aircraft of the Royal Flight is not available. Since the BAe146 of the Royal Flight is not a commercial aircraft, then its utilization will be quite small. Thus it is assumed that each of the four aircraft operates for only 5 hours/week on average (and even this may be excessive) amounting to say 250 operating hours/year. At an average flight time of say 5 hours, then an aircraft will make on average 50 flights per year. It was reported following the accident that changing the MCDPs in all four engines is carried out every 50 operating hours. This amounts to $250/50 = 5$ procedures per year on each aircraft. The probability of the omission to fit an 'O' ring is, one must assume, extremely small. This is because, if the procedure is followed properly:

● the 'O' ring is included with the maintenance kit issued by the stores,
● the person carrying out the procedure is qualified, trained and experienced in the task,
● the procedure is subject to strict quality controls including checking and signing off by a second person.

However, in this particular incident, none of the above conditions were in place and as a result events conspired together to make the omission much more likely. The frequency of such a situation occurring is not known, although it is possible that it could have been a routine violation if a long period of staff shortages had prevailed. Nevertheless, for the purposes of constructing the above fault tree, it will be assumed that the probability of such a combination of events is indeed quite low at say 1 event in 10,000 procedures or $P_{fm} = 1.0E-04$. At five procedures per year, the error will occur with a probability of $5.0E-04$/year. If the aircraft makes 50 flights per year, then the probability of this failure occurring per flight is $P_{fm} = 5.0E-04/50 = 1.0E-05$. This is the probability to be used in the fault tree.

The independent probability of failure of a major engine failure in flight requiring shutdown is not known with any certainty for this type of aircraft. However, for the purposes of the fault tree it is assumed that the probability of an engine failure during a flight is 1 in 10,000 or $P_{fe} = 1.0E-04$ per flight. The joint probability of failure of all four engines during a flight, using the multiplication rule, is therefore approximately $(1.0E-04)^4$ or a probability of $P_{f4e} = 1.0E-16$. This is exceptionally small but it must be remembered that the calculation assumes that the four simultaneous failures are due to independent causes. In practice, some common cause mechanism will almost certainly intervene and prevent this sort of extremely low probability being achieved. Such a common cause failure mechanism might be the omission of the 'O' rings when fitting the MCDPs, a failure which will effectively cause a shutdown of all four engines simultaneously, or at least within a few minutes of each other.

The scenario is represented by means of the fault tree shown in Figure 9.1. The probabilities of the four independent engine failure events, all assumed to be the same at $P_{fe} = 1.0E-04$ are fed through an AND gate resulting in a joint probability of $P_{f4e} = 1.0E-16$. This would indeed be the probability of failure of all four engines if no other external events were to intervene. However, the failure to insert the 'O' rings is an external event which can fail all four engines simultaneously. This, together with the joint probability of engine failure of $1.0E-16$ is therefore fed through an OR gate to give P_{ft}, the overall probability of failure of four engines from all causes (i.e. those considered here). Using the method of calculation for an OR gate described in Section 5.2.3, then:

$$P_{ft} = (P_{f4e} + P_{fm}) - (P_{f4e} \times P_{fm}) = (1.0E-16 + 1.0E-05)$$
$$- (1.0E-16 \times 1.0E-05) \sim 1.0E-05$$

Since P_{fm} is much greater than P_{f4e}, and because the term $(P_{f4e} \times P_{fm})$ is extremely small, P_{ft} is approximately $1.0E-05$ and therefore dominates the overall failure probability. It is equivalent to one incident of this type per aircraft every 100,000 years, which, if there were 1000 aircraft in existence, would occur at an average interval of about 100 years. The low probability of the simultaneous failure of all four engines of $1.0E-16$ in comparison with the dependent error failure therefore becomes irrelevant. The analysis has been simplified in order to demonstrate the principles involved, and is not intended to represent the real situation for any particular type of aircraft.

9.3 The railway accident at Hatfield

9.3.1 Introduction

The derailment of a high-speed train at Hatfield on the East Coast main line in October 2000, which resulted in the deaths of four passengers, was caused by the disintegration of a section of rail due to a phenomenon called 'gauge corner cracking'. It was one in a series of six major fatal rail accidents since 1997. These accidents have resulted in a total of 60 deaths and have raised serious concerns about the safety of the rail network following privatization. It was clear, right from the start, that this particular accident resulted from a failure properly to inspect and maintain the track, an activity which under privatization had been delegated to private companies. Details surrounding the direct cause of the accident, that is, the human errors involved in the failure to discover such a serious defect, are not yet available. At the time of writing the final report of the investigation by HMRI had yet to be published. The root causes of this accident have, however, been subject to great scrutiny by HMRI, the Rail Regulator and various government committees. The problem was perceived to arise from the organizational arrangements for track maintenance set up after privatization. It is mainly this aspect of the accident that is discussed below. This case study therefore comprises a more detailed examination of the organizational and management issues linked to privatization discussed in Chapter 7.

9.3.2 The accident

On 17 October 2000, an Intercity 225 passenger express train from London Kings Cross to Leeds was derailed on a section of curved track between Welham Green and Hatfield in Hertfordshire about 17 miles north of Kings Cross station. The accident resulted in the deaths of four passengers and injury to 34 others. The train was operated by Great North Eastern Railway (GNER) and at the time of the accident it was travelling at approximately 115 m.p.h. On board the train were 10 GNER staff and around 100 passengers. The locomotive and the first two passenger coaches stayed on the track, but the eight rear coaches were derailed. After the accident, six carriages stayed upright, two carriages were tilted over and the buffet car was on its side with the roof torn off from impact with the steel structure that supports the overhead power lines. The accident was investigated by HMRI and almost immediately it was found that a broken rail on the section of curved track had caused the derailment. Following this discovery, Railtrack imposed emergency speed restrictions across the whole of the railway network with immediate effect, causing widespread chaos across the country's rail transport system, a situation which persisted for many months. A metallurgical examination of the damaged track at Hatfield had revealed the existence of a phenomenon known as *gauge corner cracking*. The objective of the widespread emergency speed restrictions was to identify and renew or repair any other sections of track found to be affected by this form of damage.

9.3.3 The causes

Gauge corner cracking is a form of stress induced metal fatigue. Although well known within the rail industry, the phenomenon had not previously come to the attention of the public or media. It occurs at the point of contact between the wheels of the train and the rail. The phenomenon depends upon how much traffic is using the line, as well as the number of train wheels, the speed and condition of the wheels and the weight of the train. An inspection of an affected section of rail reveals very small surface cracks on or near the inside (or gauge) corner of the rail. HMRI prefers to refer to the phenomenon by the name of 'rolling contact fatigue'. This better indicates the way in which the damage is caused, and the definition does not restrict its occurrence to the inside of the rail.

The main defence against gauge corner cracking is first of all routine track inspections by patrol staff. Such inspections can reveal small visible surface cracks, and even cracks which are invisible can be detected by routine ultrasonic testing. Specialized equipment is available to carry out these checks. Before privatization, the inspection and renewal work was undertaken by direct employees of British Rail. Following privatization, the work was contracted out to a private maintenance contractor. When gauge corner cracking is detected, a judgment must be made to determine whether the cracked section of rail should be immediately replaced. This judgment is in turn determined by an estimation of the length of time the track can remain in operation without becoming a risk to rail safety. To remove a section of track for immediate replacement, at short notice, can be extremely costly and inconvenient in terms of service disruption. On the privatized railway, the infrastructure controller would not only lose track access fees from the train operating companies, but may also be required to pay out penalties for cancelled or delayed train services. However, if the damage is not corrected and the rail deteriorates, the cracks may extend and deepen, thus weakening the rail, eventually causing it to break. A derailment will then almost certainly follow. This may be a minor event if the train is not travelling at high speed, but if the section of track is on a high-speed passenger line then, as occurred at Hatfield, the result can be catastrophic.

If gauge corner cracking is discovered, it is not always necessary to close the track immediately. It is possible to remove cracks to some degree by mechanically grinding the rail, and in this way the life of the section of rail can be extended. However, as stated above, this is a matter of judgment, and it is important that the condition of the rail is noted and it is kept under special observation, pending permanent replacement. On the privatized railway, there was perhaps much greater economic pressure to keep lines in operation for as long as possible so that repairs could be planned at a time when disruption would be minimized.

9.3.4 Investigations

Following the accident, an investigation was carried out by the HMRI. The investigation was not restricted to track condition, but extended to an examination of the rolling stock and a review of how the train was being driven at the time of the accident. From this, it was determined that the accident had not been caused by any deficiency in the rolling stock nor by any actions of the train driver. However, in spite of some criticism of the ultrasonic methods in use for detecting cracks, it was determined that the techniques which were being used should have picked up the extremely serious condition of the rail. At the time of writing, the investigation was still not complete, and was due to examine more closely the detailed causes of the accident, in particular the management of maintenance of the track at Hatfield.

Two interim reports into the Hatfield accident have so far been published (Health and Safety Executive, 2000). In addition HMRI published in August 2002, some interim recommendations (HSE Investigation Board into the Hatfield Derailment, 2002). This provides an indication of the direct and root causes of this maintenance related accident. While it was clear that the direct cause of the accident was the fracture and disintegration of a rail as a high-speed train passed over it, the root causes appear to be linked to a failure to undertake proper maintenance of the infrastructure following privatization. The degree to which these sorts of issues are directly linked to privatization has already been discussed in Chapter 7. This accident provides yet another illustration of how cultural and management errors at the upper levels of an organization may be propagated downwards through an organization to be revealed when an accident occurs. This theme is further developed in Section 9.3.5.2

Publication of a final report of the accident by the HSE has, at the time of writing, been delayed for legal reasons, so definitive conclusions about the causes of the accident cannot be drawn at this stage. However, the seriousness of the incident was underlined by the decision of Railtrack, following the accident and under pressure from the HSE and the Rail Regulator, to undertake a rigorous survey of the entire network to identify any other defective sections of rail and where necessary, impose speed restrictions prior to rectifying faults or replacing the defective track. Numerous track problems were revealed, and as a result of the ensuing speed restrictions, and the need to take many sections of track out of service for repair, the entire UK rail network was comprehensively disrupted for a period of many months.

In August 2001, the Office of the Rail Regulator and HMRI commissioned an external report from the Transportation Technology Centre Inc. (TTCI) of Pueblo, Colorado, to examine the problem of broken rails experienced by Railtrack in the context of wider international experience. It was found that over a period of 30 years from 1969 to 2000, the number of broken rails had remained relatively constant at 700–800 per year, but then had started to increase considerably during the 1990s in the years following privatization. It was also found that over the same period the number of defective rails removed per year (as opposed to those found to be defective) had increased by a massive 600 per cent. It was also found that although the rail inspection intervals used by Railtrack compared favourably with other railway systems in developed countries, the methods used to detect problems were not so far advanced. As well as suggesting that Railtrack adopted improved methods of ultrasonic testing of rails, it was also recommended in the report that Railtrack should change to a more risk based approach in identifying maintenance work to be carried out. Comments were also made in the report about Railtrack's ability to manage the interface with its contractors so that track maintenance is undertaken in a timely manner (see below).

9.3.5 Conclusions

9.3.5.1 *Concerns before the accident*

Interestingly, the accident at Hatfield, tragic though it was, must have come as no surprise to many of those responsible for rail safety outside Railtrack. The HSE report – Railway Safety Statistics Bulletin 1998/99 – was published on 12 August 1999 by HMRI and reported a 21 per cent increase in rail breakages based on Railtrack's own statistics published some months earlier. On the same day that this report was published, the Rail Regulator wrote to Railtrack, expressing grave concern about these statistics (Winsor, 1999). Railtrack had revealed in July 1999, that 937 actual broken rails had occurred against a forecast made by the company in 1998, of 600 rail breakages for the year. The Rail Regulator's concerns naturally centred on the known linkage between track quality and safety of passengers, and regarded the inability of Railtrack to meet its targets for rail breakages as a critical test of whether Railtrack were in fact complying with their licence to operate the infrastructure. In defence of its record, Railtrack had previously referred to an apparent increase in the number of wheel flats and poor vehicle maintenance as a factor behind the increasing number of broken rails, but provided no substantive evidence in support of these views. In addition, Railtrack had claimed that the effects of the considerable increase in traffic following privatization, including a much higher volume of freight traffic, had resulted in track deterioration. These were all factors outside the control of Railtrack. However, it was pointed out by the Rail Regulator in his letter, that effective stewardship of the infrastructure required an appropriate level of maintenance of the track to suit traffic levels, whatever they were. The Rail Regulator also queried whether Railtrack had, because of the claimed deficiencies in the condition of the rolling stock, taken any action to bar certain trains or classes of train from the infrastructure pending remedial action to correct those deficiencies. This apparently had not occurred. In June 2000, only a few months before the Hatfield accident, HMRI Chief Inspector of Railways wrote to Railtrack expressing concern about Railtrack's failure to meet their agreed targets. The main thrust of this correspondence was that Railtrack and their contractors needed to adopt a more co-ordinated approach to maintenance of the infrastructure in order to ensure safety.

9.3.5.2 *Management of track maintenance*

9.3.5.2.1 The impact of privatization

Before privatization of the railways, not only did responsibility for infrastructure maintenance lie with the nationalized British Rail, but also the work was actually undertaken by permanent British Rail employees on a regional basis. Due to the vertically integrated structure of the railways, if track deterioration occurred as a result of problems with the rolling stock, this was also the responsibility of British Rail, with no possibility of transferring the blame to a separate train operating company. Within the British Rail organization, there existed a huge reservoir of skilled staff that had experience in maintaining the infrastructure stretching back over many decades, supported by comprehensive data and records concerning the condition of the track.

Not only was there local detailed knowledge of the infrastructure and its condition, there was a single 'controlling mind' overseeing both infrastructure and trains.

The track maintenance performance of British Rail, pre-privatization, was not of course perfect. An accident at Hither Green in December 1967 resulted from poor track maintenance and an inadequate line speed limit. It was one of the worst post-war rail disasters in Britain and caused the deaths of 49 people when a crowded commuter train from Hastings to Charing Cross was derailed. The cause of the crash was poor maintenance, in particular the poor condition of sleepers and loose rail fastenings. The inquiry into the accident recommended, among other measures, more frequent ultrasonic testing of track and improved maintenance standards.

Following privatization, large numbers of managers and staff experienced in infrastructure engineering, were encouraged to leave the industry, either to take early retirement or to take jobs with the private companies to whom track maintenance had been sub-contracted. In addition, the direct labour force previously employed by British Rail was largely transferred *en masse* to work for these companies under new terms and conditions, many of these working through agencies on a short term, casual basis. Many of them never worked on one contract for perhaps more than a few days or weeks at a time. Although there were a number of large contracting companies tendering for work, 50 per cent of contracting firms working in the industry employed less than ten workers. This calls into question whether these small firms have the resources to provide adequate competence training. Many of the companies working on infrastructure maintenance are better known as construction companies, and it is well known that in the construction industry, some 94 per cent of companies employ less than eight workers.

Within a matter of months of privatization, the large centrally directed maintenance structures of British Rail had effectively been dismantled to be replaced with a loosely controlled *ad hoc* organization of contractors, many of whom had no previous experience in track maintenance. The nature of the contractual agreements was such that contractors would get higher fees if they completed the maintenance in less than the designated time and if they were able to keep track closures below a specified minimum. The danger of this sort of agreement is that what should be routine and meticulous work is rushed with the potential for mistakes or omissions in order to maximize income. Following the Hatfield accident, the structures that had evolved out of privatization for track maintenance were widely perceived to be the most probable root cause of the tragedy.

9.3.5.2.2 Railtrack's use of contractors

The situation whereby work was contracted out under privatization came under close scrutiny by HMRI following the Hatfield derailment. A report by the Health and Safety Commission (HSC) published in May 2002 (The Health and Safety Commission, 2002) pointed out that the responsibilities of Railtrack for health and safety were defined by the criminal law and it was not possible to transfer the responsibilities from one party to another by contract. As a precedent for this, the decision by the House of Lords in 1996 in the case of *Regina v Associated Octel Co. Ltd* was quoted, indicating that an organization always retains final responsibility for work carried out as part of their operations by contractors. Thus, although Railtrack had a legal duty under its licence conditions to ensure proper maintenance of the infrastructure, following privatization a policy was adopted whereby contractors were held entirely responsible for the quality of the work carried out, for carrying it out safely, and for meeting the Railway Group Standards (RGSs). In practice, this means that although the RGSs may be superlative in promoting the highest standards of maintenance, unless there is adequate oversight by the Duty Holder (the responsible person under the Health and Safety at Work etc. Act 1974) of the way work is carried out, then the standards themselves become irrelevant.

In a similar way, Railtrack had prepared a safety case for its operations as required under the Safety Case Regulations (The Health and Safety Executive, 2000). The HSC report (Health and Safety Commission, 2002) notes that although the safety case does set out the arrangements for selection, monitoring and control of contractors, there were many weaknesses in the systems that were supposed to enforce the principles described in the safety case. In order to address this issue, major revisions were made by Railtrack to their safety case under the guidance of an 'intervention plan' prepared by HMRI.

The HSC report also notes that although train derailments have generally decreased in number against a considerable increase in traffic since 1997, over the past 2 years, there had been three major derailments of passenger trains resulting in a total of 21 fatalities (these accidents were Hatfield, Potters Bar (see below) and Great Heck (caused by train impact with a road vehicle)). In the 6 years before Hatfield there had only been one derailment resulting in fatalities (two deaths at Branchton) and this had been due to vandalism. The report examines the qualifications and training of track workers employed by the contractors of Railtrack. The main qualification is the Personal Track Safety (PTS) certificate. This, valuable as it may be, is only designed to protect workers from death or injury while working on the track and is not intended to measure the competence of contractors' employees to carry out the work adequately. The report makes the important point that Railtrack, as the Duty Holder, did not have any requirements or system of formal qualifications to ensure the competence of track maintenance workers. However, by the time of publication of the report, schemes were being prepared by the appropriate industry bodies to provide standards and training modules for track maintenance and renewals to improve the competence of workers and supervisors. Another important matter raised by the HSC report was the lack of a comprehensive and reliable asset register which described the condition of the infrastructure. Following privatissation, knowledge about the infrastructure became fragmented among the various contracting companies involved in its maintenance. This only came to light after the Hatfield accident.

Specific recommendations concerning Railtrack's management of contractors also arose out of the Cullen recommendations from the Ladbroke Grove accident inquiry (Health and Safety Executive, 2002), which also encompassed the Southall accident. Lord Justice Cullen reported that 'the evidence in regard to the use of contractors was a source of considerable concern'. In his comments he raised many of the same points regarding training and competence of the contractor workforce, management controls on the work of contractors and the need for a reduction in the number of contractors employed.

Table 9.2 Qualitative human error analysis of the Hatfield accident

Cause	Description	Type of error or other failure	Systemic cause or preventive action
Direct cause	Failure to detect and respond to evidence of rolling contact fatigue of the track resulting in disintegration and fragmentation of the rail as the train passed over.	Latent error	Inadequate management systems to ensure serious deterioration of the rail is responded to in a timely way. Poor communications and feedback between infrastructure controller and contractor in regard to need for urgent action.
Root cause	Failure to follow best practice in rail maintenance as set out in RGSs.	Violation	Lack of supervision and failure of management oversight by the infrastructure controller.
Contributory causes	Post-privatization contractual agreements for rail maintenance which reward speed with which work is carried out rather than the quality of the work.	Latent error	The cause was a dismantling of the pre-privatization arrangements for rail maintenance and their replacement with inadequately managed private contracts. Lack of knowledge of the condition of the infrastructure.

9.3.5.3 *Human error analysis*

The qualitative human error analysis of the Hatfield accident set out in Table 9.2, can only be based on interim recommendations from the various safety reports issued by HMRI, and is not therefore a definitive account of the underlying causes.

9.4 The railway accident at Potters Bar

9.4.1 Introduction

On Friday 10 May 2002 the 12.45 train from Kings Cross to King's Lynn was derailed just outside Potters Bar station, only a few miles from Hatfield. It resulted in seven fatalities, including one pedestrian who was hit by falling debris. It was found that the derailment had occurred when a set of points south of the station moved while the train was passing over them. As a result, the fourth carriage left the track and skidded sideways damaging a road bridge over the railway before coming to a halt wedged under the Potters Bar station canopy. At the time of writing the investigations into this accident were not complete and the following analysis is based on initial and interim reports by HMRI. Definitive conclusions cannot therefore be drawn and a detailed human error analysis, as presented for other case studies, is not included. However the accident is of interest occurring as it did within about 18 months of the accident at Hatfield, described above, and a few miles down the track, since initial findings indicate that the direct cause was maintenance related with potentially the same root causes.

9.4.2 Initial investigation – direct causes

It was decided by the government that a public inquiry would not be held into the accident. Instead an Investigation Board was set up under the Health and Safety at Work etc. Act 1974. The Board comprised members independent of HMRI but included two senior HSE inspectors. An investigation was carried out by HSE and an interim report was published (Health and Safety Executive, 2002). The immediate findings were that the points had moved because two pairs of nuts on the stretcher bars had become detached. After the accident they were found lying on the ballast at the side of the track. The purpose of the stretcher bars is to hold the parallel tracks the correct distance apart. When the bolts came loose, the locking bar which connects the tips of the points moved and then failed due to the abnormal forces exerted when the train passed over it. The reason why the nuts had become detached became the focus of further investigations. The track was maintained by a large private construction company, Jarvis plc, who said that safety inspections had been carried out the day before the crash and that no faults were noticed. However earlier, on 1 May, engineers had found that two nuts had become detached and were lying beside the points. They were immediately reinstated although the repair was not recorded in the maintenance log as required. In the weeks following the disaster, Jarvis plc suggested that the loosening of the nuts might have been caused by sabotage. However, later investigations by HSE and British Transport Police found no evidence for this. Jarvis plc had been awarded a £250 million contract for the maintenance of the line under a type of agreement where fees are not fixed but depend upon the speed at which the work is completed. Jarvis denies responsibility for the accident.

9.4.3 Later findings – root causes

In May 2003, a further progress report was published by HSE (HSE Investigation Board, 2003) while the British Transport Police continued to carry out a criminal investigation to determine whether a case for corporate manslaughter could be answered. The earlier report had identified the direct cause of the accident as the failure of the defective points. The latest report identifies the root cause of this failure as poor maintenance and proceeds to investigate the underlying reasons, in particular the systems for managing inspection and maintenance including issues of training, competence and supervision. It reveals that two separate systems existed for the inspection and maintenance of points, involving signalling and permanent way staff respectively, with possible confusion about roles and responsibilities. In addition there was a lack of clear guidance and instructions for inspecting and maintaining points of this type. This, coupled with reported deficiencies in the arrangements for ensuring staff competency, would seem to be a recipe for disaster. The report also notes a 'failure to recognize safety related defects in the set-up and condition of the points and to record or report them'. On the night preceding the accident, there had been a report submitted of a 'rough ride' of a train passing over the same points. If this report had been acted upon the accident that took place the following day might have been prevented.

Although no record could be found of the locking nuts being removed or replaced, other similar points in the area were inspected and a wider problem of locking nuts working loose was indicated. On some points it was found that friction washers had been fitted and on others evidence was found of 'pop' marks, threads indented with a hammer and cold chisel to seize the thread and stop nuts working loose. This is bad engineering practice, certainly for a safety critical item, and prevents further adjustments being made. It would seem to indicate inadequate management oversight of work practices. In later discussions, Railtrack indicated that it accepted nuts and bolts would come loose and when these were discovered they would be tightened up during maintenance. In the meantime safety would be ensured by relying on other stretcher bars in the set of points. The later investigations found no evidence that the locking nuts on the incident points came loose as a result of sabotage.

9.4.3.1 *Conclusions*

Based on the HSE investigation, the Investigation Board made a total of 26 recommendations. It concluded that the points were 'not "fit for purpose" for the operating environment and safety related functions expected of them'. In addition an investigation across the network (by Railtrack) found that 48 per cent of 870 sets of points examined deviated from the 'normal' condition, and although none matched the condition of the Potters Bar set of points, all were found to have one or two deficiencies per set. As the investigation is still ongoing and a criminal investigation is in progress, no specific details of human errors which actually took place during maintenance procedures have been identified. However, the accident at Potters Bar served to raise further concerns about the way maintenance is contracted out under the post-privatization arrangements.

9.4.4 Postscript – the future of rail maintenance?

On 14 May 2003, it was announced by Network Rail, the not-for-profit company that replaced Railtrack in October 2002, that it intends to take away control of maintenance on the high-speed lines from Reading to Paddington from a private contractor, Amey plc, and undertake its own repairs and inspections. It is intended that all contract staff currently working on the line will transfer to Network Rail, a reversal of the post-privatization arrangements. The reason given was that Network Rail wished to retain direct control of maintenance and thereby gain a better understanding of the condition of the infrastructure. This return to the traditional way of conducting infrastructure maintenance was said to have been prompted by the initial findings of the investigation into the Potters Bar accident. In particular there was concern that the contractor had been unable to offer an explanation of why the locking nuts on the points had become detached. It has also been reported that Network Rail has instructed all its maintenance contractors to increase the number of permanently employed maintenance staff from the current 50 per cent at present to 85 per cent by the end of 2003, thus reducing their reliance on casual labour supplied by agencies. Cost considerations had also entered into the decision, since as a result of the Hatfield and Potters Bar accidents, risk insurance premiums, to protect contractors against claims for death and injury resulting from accidents, were being added to the price of maintenance contracts making them prohibitively expensive. It seems probable that the new policy may become more widely applied across the network.

References

HSE Investigation Board into the Hatfield Derailment, (2002). www.hse.gov.uk/railway/hatfield/investigationb1.pdf. Interim Recommendations of the Investigation Board.

HSE Investigation Board, (2003). www.hse.gov.uk/railways/pottersbar/may03progrep.pdf. Train Derailment at Potters Bar 10 May 2002 – a Progress Report May 2003.

The Health and Safety Commission, (2002). *The Use of Contractors In the Maintenance of the Mainline Railway Infrastructure,* London: The Stationery Office.

The Health and Safety Executive, (2000). *Railways (Safety Case) Regulations 2000, Guidance on Regulations L52*, Sudbury, Suffolk: HSE Books.

The Health and Safety Executive, (2000). *Train Derailment at Hatfield, 17 October 2000, First and Second HSE Interim Reports*, Sudbury, Suffolk: HSE Books.

The Health and Safety Executive, (2002). *The Ladbroke Grove Rail Inquiry*, (Parts 1 and 2), Sudbury, Suffolk: HSE Books.

The Health and Safety Executive, (2002). *Train Derailment at Potters Bar, Friday 10th May 2002*, Sudbury, Suffolk: HSE Books.

Winsor, T. (1999). www.rail-reg.gov.uk/pn99/pn50_99.htm. A formal requirement from the Rail Regulator for further information about Railtrack's plans for tackling the problem of broken rails and for confirmation of what those plans are intended to achieve.

10

Active errors in railway operations

10.1 Introduction

The safety of railway operations has always depended upon the reliability of train drivers, signalmen, maintenance staff and other personnel. On the railways there has always been a strong link between a single active human error and a major accident. Since the inception of the industry in the early 19th century railway engineers have attempted to break that link. This has usually been done by installing electro-mechanical devices and interlocks between the human and the system, varying from the simple to the ingenious. Many of these devices attempt to avert a train collision by preventing two trains occupying the same section of track. The continuing quest to break the link between error and consequence has mainly been driven by the occurrence of major accidents involving loss of life. Each new accident revealed a new loophole in the existing safety system, usually one that involved a previously unforeseen human error.

Until quite recently, the approach to improving railway safety had changed little since the 19th century. The traditional approach was that the accident would be investigated by Her Majesty's Railway Inspectorate (HMRI) and recommendations made to prevent the same accident occurring elsewhere. The recommendations would be adopted across the industry and enshrined in standard rules and regulations published as the 'Rule Book'. All staff are issued with this and expected to obey it to the letter. This approach had two major effects:

1. The railways evolved into a pseudo-military organization, where safety depended upon orders, rules and regulations being obeyed largely without question. This system was upheld by HMRI who, up until the 1990s, recruited its staff almost entirely from retired officers of the Royal Engineers. These were men with long practical experience in engineering and with a military rule-based background. The system was in many ways highly effective in promoting railway safety over a long period of time.

2. Safety improvement measures had to wait for an accident and were directed towards preventing the same or a similar accident happening again. The measures were then widely implemented across the railways, in the early years, through Acts of Parliament. While beneficial, this approach tended to address the direct cause rather than the root cause of accidents. Since the root cause had not been eliminated, it was almost guaranteed that a similar accident would soon occur with a different direct cause. The end result was that accidents continued to happen and resources were wasted in trying to plug the loopholes.

The traditional rule-based, accident driven approach to safety has not been limited to the railway industry. It is still widespread and is often referred to as the 'prescriptive' approach to safety. While the approach has some value the principal drawback is its reactive nature. It tends only to identify and respond to new hazards after an accident has occurred. Today a more proactive approach is adopted by the railway industry enforced by wide-ranging legislation enacted following a major rail accident at Clapham Junction in 1987. The new legislation for the railways was largely based on regulations imposed on the offshore oil and gas industry following the Piper Alpha fire and explosion. The proactive approach has not, however, completely replaced the prescriptive or rulebook approach, nor should it, since the two approaches can be entirely complementary if the correct balance is maintained.

This chapter examines the way in which the prescriptive approach was developed on the railways and in many respects came to be highly successful in controlling the more extreme consequences of active human errors in railway operations. It goes on to examine current approaches to these human errors using a number of recent railway accidents as case studies to illustrate how a number of common, yet seemingly intractable, systemic human error problems are being resolved. As train collisions are the main type of accident considered here, it is necessary to include a description of the various systems, both old and new, which are used to ensure train separation. These are described in the Appendix, which should be read in conjunction with the following sections.

10.2 Signals passed at danger

10.2.1 Introduction

Modern signalling systems are extensively automated, extremely reliable and are based on a fail-safe principle, such that if a fault occurs, the system reverts to a condition whereby all train movements are halted. A notable exception to this was the major accident at Clapham Junction in 1988, briefly described below, which was due to a signal wrongly indicating a 'proceed' aspect as a result of an electrical circuit fault in the signal control equipment. This accident illustrated that although modern signalling systems are extremely reliable they are still vulnerable to dangerous latent errors in maintenance. More importantly, in spite of signalling improvements, safety is still highly reliant on train drivers observing and responding correctly to every caution and danger signal encountered on route. When, as sometimes happens, a train driver passes a signal showing a red or danger aspect, for whatever reason, this is technically known within the industry as a signal passed at danger (SPAD). The historical problem of two trains coming to occupy the same section of track has not therefore been completely solved. The SPAD falls into the category of an active error because its effects are immediate, given that the train overruns a red signal and enters a section occupied by another train.

The error resulting in the false signal at Clapham Junction was a latent error during maintenance, since a number of hours had elapsed between the error and its consequence. In this sense the accident could more appropriately be located in the preceding chapter. It is briefly included below because it was an important milestone in the evolution of safety on the railways and marked a change from a mainly prescriptive approach to safety, to one that is more proactive and hazard based and therefore more in line with modern accident prevention techniques. More relevant to the study of active errors are the fatal railway accidents at Purley, Southall and Ladbroke Grove included later in this chapter.

10.2.2 The accident at Clapham Junction

10.2.2.1 *Introduction*

Changes from the traditional to more modern methods of signalling and train control are described in the Appendix. These changes have virtually eliminated the human operator from the track side of railway operations. As a result, train accidents due to human errors by signalling or train control staff are now extremely rare. The most recent serious accident in recent years to result from signalling errors occurred at the approach to Clapham Junction station in London in December 1988. A rear end collision occurred between two commuter trains approaching the station on the main line to Waterloo from the South-west of England. It occurred not as a result of active human error in the operation of the signals, since these were operated automatically, but was due to a latent maintenance error.

10.2.2.1.1 The accident

The 06.14 Poole to Waterloo train en route to central London was presented with a series of green signals from Wimbledon. Two signals out from Clapham Junction the driver passed another green signal, one carrying the code number WF 138. The signal was located just before a bend in the track which restricted the forward vision of train drivers entering that section. On passing the green signal and rounding the bend in the track, the driver saw a stationary train ahead of him but was unable to stop in time and ran into the rear carriage at a speed of 35 m.p.h. Another train coming in the opposite direction on the adjoining track then ran into the wreckage of the first two trains which was strewn across both tracks. A train following behind the Poole train then passed through the same signal WF 138, which had been green, but was now showing a single yellow, meaning proceed with caution. The section ahead was now occupied by the two trains which had collided. Fortunately the driver was able to see the rear of the wrecked Poole train and applied his brakes in time to prevent another collision. The driver of the Poole train and 32 passengers in the trains involved in the disaster were killed instantly and two more passengers died later.

10.2.2.1.2 The inquiry

The Public Inquiry into the accident was conducted by Mr Anthony Hidden QC (Anthony Hidden QC, 1989) and was crucial in bringing about major changes to the way railway safety was managed and regulated in the UK. One of the more important outcomes was that HMRI ceased to be an organization within the Department of Transport, where it was believed there could be a conflict of interest. It was instead transferred to within the aegis of the Health and Safety Executive (HSE) where it would be exposed to more modern methods of safety regulation. At the same time, HMRI ceased to recruit its inspectors from the Royal Engineers, and instead began to recruit health and safety specialists.

10.2.2.1.3 The causes

The inquiry found that the cause of the accident was a signal fault resulting from weekend working in the main signal box controlling trains at Clapham Junction. The work involved installing a modern system of electrical relays. This involved isolating and removing the old relay boxes, but ensuring the signalling system would be operational for the Monday morning rush hour. In carrying out the work, a technician had made a very simple but catastrophic error when he failed properly to insulate the live ends of two electrical wires which had been disconnected from an old relay box. The result was that the two wires came into intermittent contact causing false electrical currents to be sent to signal WF 138 such that it began to operate erratically, changing aspects irrespective of whether a train was present in the block that it was supposed to be protecting. This was a latent error that occurred because of a lack of procedures setting out best working practice and which had remained undiscovered because of inadequate checking and quality control systems. In addition the technician had been working extremely long hours unsupervised. The result was that all the conditions for a catastrophic latent error were in place.

Prior to the accident, a number of other train drivers had noticed unusual aspects that morning on signals on the track leading into Clapham Junction. The driver of the train ahead of the Poole train had in fact seen a green signal as he approached but in the last few seconds just before he passed the signal it changed suddenly to red. He had therefore brought his train to a halt in the section in order to call the signal box assuming that his train would be protected to the rear by red and yellow signals. However, the driver of the Poole train on seeing a green signal, had driven into his rear.

10.2.2.1.4 Conclusion

The Hidden Inquiry into the Clapham Junction accident produced 93 recommendations relating to improved methods of conducting and supervising maintenance and the adoption of quality control techniques to ensure that work has been carried out properly. These recommendations were applied across the whole of the railway network, including non-British Rail services such as Underground Trains and Light Railways. The inquiry also strongly recommended that British Rail improve its overall safety management systems and included a recommendation to install an automatic train protection (ATP) system across the network to prevent signals being passed at danger. Over 10 years later, this recommendation had still not been implemented when a series of SPADs resulted in two high profile multiple fatality accidents at Southall and Ladbrook Grove. However, prior to examining these accidents it is useful to examine an accident which occurred at Purley in South London in 1989. This accident raised important issues about driver responsibility for committing SPADs and the attribution of blame.

10.3 The train accident at Purley

10.3.1 Introduction

One of the most serious rail accidents to occur in the UK in recent decades was caused by a SPAD and took place at Harrow in October 1952 causing 122 fatalities. It occurred when the Perth to Euston sleeping car express passed a series of signals at danger in fog and ran into the back of a local commuter train standing at a platform in Harrow station with 800 passengers on board. In this accident, as in many serious train collisions, the driver of the express train was killed in the accident. Thus, although it was known how the accident occurred, the investigators were never able to establish why. The issue of SPADs remained unresolved and although similar accidents continued to occur at regular intervals it was only following an accident at Purley, South London in 1989 that the issue attracted public attention. This was probably because the accident at Clapham Junction 3 months earlier was still fresh in the public memory, leaving serious concerns about railway safety.

10.3.2 The accident

The accident at Purley on the London to Brighton line about 13 miles south of Central London involved two passenger trains and occurred on Saturday 4 March 1989. The 12.50 Horsham to London train, a four-car electrical multiple unit (EMU), was passing from a slow line to a fast up line and was hit by an express train travelling from Littlehampton to London Victoria. The driver of the express had passed through a series of signals at cautionary or danger aspects. These were:

- a double yellow aspect signal (the first warning to proceed with care because of a train two sections ahead),
- a single yellow aspect signal (the second warning to proceed with care due to a train one section ahead),
- a red aspect signal (a train in the next section).

The express ran into the back of the slow train as a speed of about 50 m.p.h. Five people were killed and 88 passengers were taken to hospital. Carriages from the express train left the track and overturned, rolling down a steep embankment coming to rest in the gardens of adjoining houses. The driver survived the accident only to be prosecuted a year later for manslaughter and jailed for 18 months.

10.3.3 The inquiry

An inquiry into the accident was chaired by the Deputy Chief Inspector of Railways. It was found that on the day of the accident the signals had appeared to work correctly but that the driver of the express had passed through two cautionary and one danger signal, after having unconsciously reset the automatic warning system (AWS) alarms. The AWS, fitted to all UK trains to guard against SPADs, is described in detail in the Appendix.

During the inquiry into the accident, it was revealed that SPADs were a common occurrence on the network. The charge of manslaughter against the driver and the jail sentence which followed was unusual. It is possible that legal proceedings were taken to allay public anxiety by demonstrating that the safety issue was being taken seriously. While driver responsibility is important in avoiding SPADs, the AWS has serious short-comings and does not take account of the way human errors occur (see Section 10.4 below). As a result, the system leaves both drivers and passengers vulnerable to the consequences of active errors from which recovery is a matter of chance. Since these errors are systemic, the problem of SPADs is not best dealt with by punishing the train driver but by addressing the system faults.

10.3.4 Conclusion

The direct cause of the accident was the failure of the driver to respond to a series of signals. However, the root cause of the accident was not the driver's loss of vigilance, or his violation of the rules (passing a signal at danger), but that the AWS intended to protect him is inadequate for its intended purpose. Prosecuting the driver for manslaughter did not address the real safety issue and enabled the systemic cause to be ignored. As a result it was inevitable that more accidents would take place due to SPADs. Following privatization, two multiple fatality train accidents due to SPADs occurred within a period of 2 years at Southall (1997) and Ladbroke Grove (1999). It was only following these accidents, and the ensuing public concerns about railway safety, that the issue of SPADs was taken more seriously. The problems inherent in the AWS are discussed below prior to the case studies of the two accidents.

10.4 The driver's automatic warning system

10.4.1 System operation

A technical description of the driver's AWS and other train protection systems can be found in the Appendix. The AWS is a driver vigilance device and not a system for automatically controlling the movement of a train. It monitors the vigilance of the driver and has the capability to stop a train if he fails to acknowledge a warning bell and visual indicator in his cab, by pressing a plunger. These warnings are activated whenever a yellow or red signal is approached. If the signals are green there is no activation of the AWS. When the warnings are activated the driver must depress the plunger to avoid an automatic application of the brakes. If a green signal is encountered the driver hears a bell with a different sound. The following description of the practical operation of the system is based on four aspect signalling.

If the warning system alerts the driver to a cautionary double yellow aspect signal (meaning there is a train two sections ahead) then the driver will immediately cancel the AWS. He may need to apply the brakes manually to slow the train to what he considers to be a safe speed to enter the next section. Whether braking is needed and how much, depends on the speed of the train, its weight, its braking characteristics and the safe speed for the section ahead. The driver will now be aware that the next signal could be at single yellow (meaning there is now a train one section ahead). This would indicate that the train in front is still in the same section as when he passed the double yellow signal and therefore may be stationary or slow moving. Thus he may be required to make another speed reduction. Alternatively the next signal may be double yellow (train has moved on and is still two sections ahead) or it may be green if the train has vacated the section ahead. It will not be red. A red signal is always preceded by a single yellow signal. Depending on the speed of his train and his knowledge of the route, it may not always be necessary to reduce speed and the driver may decide to postpone the braking action. If the AWS warning sounds again and he sees a red signal he is required to stop because there is a train in the next section. The previous double and single yellow aspect signals will have already given him sufficient warning to reduce speed so that he is able to stop without overrunning into the next section. Usually some safe distance is provided after the red signal and before the next section starts and this is referred to as 'overlap'.

This is how the system is supposed to operate, but it does have two particular dangers which are described below.

10.4.2 A conditioned response

During the course of a journey a train driver will encounter numerous AWS warnings which he is required to cancel. The design of the system is based on the assumption that the driver will only cancel the warning when he has noted the yellow or red signal. If he fails to note the yellow or red signal then the audio–visual warning will alert him to its presence. However, because he responds to so many warnings, the cancelling process can become a conditioned response. A bell is heard, an indicator changes colour and a plunger is pressed. This is more probable if the driver's attention has drifted away or he is distracted by other events, for instance, having to speak to the guard. The driver has cancelled the first AWS alarm but has taken no action. If the driver has taken no action, then it may be because he has decided no action was necessary. The speed of the train may not have necessitated a further speed reduction or possibly he has decided to postpone the speed reduction. If this is the case, at least he has taken heed of the double yellow signal. However, it is also possible that no action was taken because the driver made a conditioned response to the warning by cancelling the alarm.

If the next signal has a single yellow aspect and he sees it, then there may still be time available for the driver to make the necessary speed adjustment. At the single yellow signal, the AWS will sound again and reinforce the need for action. This aspect indicates that the headway between the two trains is decreasing and the train in front is now only one section ahead. The second AWS will be cancelled and a further speed reduction will almost certainly be necessary since the next signal to be encountered is likely to be a red danger signal. However, if this second cancelling action is also a conditioned response then no speed reduction will be made. There is now much less time for the driver to recover from the omission to take action at the double yellow and single yellow aspect signals. If the next signal is a red signal, then the AWS alarm bell will sound and will need to be cancelled. The train must now be brought to a standstill, since the train in front is now clearly stationary in the section ahead. If this final cancellation is a conditioned response, the driver may still avoid an accident if he observes the train ahead and is able to stop in time. At this late stage it will depend entirely on the speed of the train, its weight and its braking characteristics whether a collision can be avoided. It may be possible to stop the train before overrunning the section containing the train in front if the speed is low enough. If the speed of the train is too great, then there will be an overrun, the train will enter the next section, and depending on the overlap provided, there may be a collision.

It may seem highly improbable that a driver could pass through two yellows and a red signal, without responding, having successively cancelled a series of AWS alarms. The case studies of rail accidents caused by SPADs described in this chapter, as well as numerous accidents not mentioned here, testify to the sad fact that it is entirely possible. In fact many hundreds of SPADs of exactly this type are reported every year where, fortunately, no accident has been caused (see Section 10.6.2). The chance of further signals being passed after having passed an initial cautionary signal may well be increased due the effects of dependency described below.

10.4.3 The effect of dependency

In order for a driver to pass through a red signal, he must have already passed through a series of double yellow and single yellow cautionary signals. In addition, these must have been acknowledged by cancelling the AWS or else the train would have been stopped automatically. As described above, cancelling the AWS alarm might become a conditioned response but because this is a sequence of similar tasks close together in time and location, there is also the possibility of human error dependency. As discussed in Section 5.4 of Chapter 5, once an initial error has been made, the probability of subsequent errors is increased where there is some form of coupling between the tasks. In this case it is possible that the dependency between errors will be 'internally induced' where the coupling is that the same individual carries out similar tasks close together in time or place. Alternatively the dependency could be 'externally induced' by, for instance, the need to respond to the driver's radio or to deal with some event on the train brought to his attention by the guard. There are clearly many other possibilities. Whatever the coupling mechanism involved, dependency will almost certainly increase the probability of a SPAD once the first cautionary signal cancellation has been made with no response by the driver.

The recent accidents at Southall and Ladbroke Grove brought the problem of SPADs to public attention and show a SPAD can lead to a serious accident when circumstances work against error recovery.

10.5 The Southall and Ladbroke Grove rail accidents

10.5.1 Introduction

This section describes the two major accidents due to drivers passing signals at danger which have occurred since privatization in 1997. The first of these accidents, which took place at Southall, London in 1997 caused six fatalities and 150 injuries. A Public Inquiry was set up but was delayed by 2 years due to criminal proceedings against the driver and corporate manslaughter charges against the railway company. By the time the inquiry commenced, another accident had occurred at Ladbroke Grove only a few miles along the track, again as a result of a driver passing signals at danger. A second inquiry was then opened with a requirement to learn lessons from both accidents. Each accident is described below.

10.5.2 The Southall rail accident 1997

10.5.2.1 *The accident*

On 19 September 1997 the 10.32 high-speed passenger train from Swansea to London operated by Great Western Trains went through a red danger signal at Southall about 9 miles from Paddington station. It collided with an empty freight train operated by English Welsh and Scottish (EWS) Railways that was crossing the path of the passenger train. The accident resulted in six fatalities and 150 injuries. It occurred on a busy section of track where there were four main lines joined by crossovers between the adjacent tracks.

There was no problem with the operation of the signals which allowed the EWS freight train, comprising empty hopper wagons, to cross over from a relief line to some sidings on the south side of the main line. When the movement occurred, the sections of the main lines which were temporarily occupied by the freight train were fully protected by interlocking between the crossover points and red signals on both the up and down main lines. However, before the freight train was able to clear the main line, the high-speed passenger train ran through the red signal protecting the section at a speed of 100 m.p.h. Prior to this it had run through a set of cautionary yellow signals in turn. The power car of the passenger train struck the empty hopper wagons a glancing blow and came to a standstill about 100 yards down the line. The first coach overturned but the second coach impacted an electrification mast and was bent almost double. Most of the seven fatalities occurred in this coach.

It was discovered that the train had been fitted not only with the normal AWS but also with an automatic train protection system (ATP) which had been operating on a trial basis. On the day of the accident, neither of the systems were operational so the driver and passengers were completely unprotected from the possibility of a SPAD. Even so, it was decided that the driver of the train should be prosecuted for manslaughter.

10.5.2.2 *The inquiry*
10.5.2.2.1 The prosecutions
A Public Inquiry into the accident was ordered by the government to be chaired by Professor John Uff and this was set up within a few days (John Uff, 2000). However it was then subject to a 2-year delay for legal reasons associated with criminal prosecutions against the driver and Great Western Trains. It was believed that if staff were questioned at a Public Inquiry, then the evidence that was given might prejudice any later legal proceedings to prosecute those responsible. This was unusual, because the normal course of events in the event of an accident, would be to proceed with the inquiry but not publish the report until the Director of Public Prosecutions had decided whether a prosecution was in order. On this occasion, for unknown reasons, normal precedent was not followed.

Following the Southall accident, the driver of the passenger train was fortunate enough to survive but was questioned by police almost immediately after the accident and later charged with manslaughter. In addition to the charges made against the driver, charges of corporate manslaughter were brought against Great Western Trains. However, both charges were dropped some 20 months later allowing the accident inquiry to commence. The corporate manslaughter charges were abandoned because no one individual responsible for disabling the protection systems could be identified. In a pre-trial hearing, it was ruled by the judge, Mr Justice Scott-Baker, that prosecution could only take place if a person deemed to be a 'controlling mind' could be identified and charged. In the end no such person could be found. Perversely, the manslaughter charges against the driver were only dropped because it was judged he was unfit to plead and would be unable to pay a fine, rather than because of his obvious vulnerability to human error through the disabling of both protection systems. Great Western Trains were also charged under the Health and Safety at Work Act for failing to protect the public to which charge they pleaded guilty and were fined £1.5 million.

10.5.2.2.2 The inquiry findings
Hearings finally took place between 20 September and 25 November 1999. It was revealed at the inquiry that the train involved in the Southall accident had been fitted with two types of train protection system, but both of these had been disabled with the implicit approval of management. One of the systems was the ATP system (see Appendix). This was a system which the Hidden Inquiry, a decade earlier, had recommended to be fitted to all trains. The ATP system in use on the train at Southall was

a prototype which had been on trial but was not fully operational due to unreliability. At that time very few trains had been fitted with this system. The other system which had been disabled was the AWS which is fitted to all UK trains (see Appendix and Section 10.4).

It was found that the rules for allowing a train to operate without the AWS operational were ambiguous and over the years had evolved considerably. British Rail had introduced the AWS during the 1960s and at that time it was the general policy that AWS was only an aid to the driver and did not remove the ultimate responsibility for responding to signals. In the 1960s not all trains had been fitted with AWS. This policy was construed to mean that a train would be allowed to operate without the AWS. Eventually, the rules were modified so that if the AWS on a train failed, then the train had to be taken out of service for repairs as soon as practicable. Later changes to the rules modified the wording to the effect that the train had to be taken out of service at the first suitable location without causing delay or cancellation. This was the rule in force at the time of the accident at Southall.

By the time of Southall, the wording of the rules was so ambiguous it was interpreted to mean that a train without a functional AWS might be able to complete its journey, if it were deemed that a suitable location did not exist along the route. Thus an immediate cessation of running would not be necessary minimizing disruption to the service. This also meant that the driver might be required to complete a long journey responding visually to numerous cautionary and danger signals along the route. There had been a rule, at some earlier time, that the train would only be allowed to operate without AWS when a second person was present in the cab. However there was concern that this could pose a greater risk than not having AWS working, because of the potential for distraction of the driver. The only qualification required for the second person was that he was not colour blind! Following privatization of the railways, and cuts in railway staff, it became less likely that a second person would be available at short notice.

The inquiry also found that the driver's view of the signals in the Southall area was impeded by work on overhead line electrification. This had involved raising the height of many of the signals to as much as 18 feet above the ground making them more difficult for the driver to see. The first double yellow signal which the driver had passed was obstructed by an over-bridge at Southall station. It was also found that the next single yellow signal had not been correctly focused and the beam was only in the driver's sight for about one second at the speed of the train. By the time the train had reached the red signal, having failed to slow down for the cautionary signals, it was travelling at too high a speed to stop before it collided with the freight train in its path.

10.5.2.3 *Conclusion*

The direct cause of the accident was the failure of the driver of the high-speed train (HST) to respond to a series of cautionary and danger signals, allowing the train to overrun into a section occupied by the freight train crossing the mainline. However, the root cause of the accident was almost certainly an operational rule which allowed the train to be driven with both train protection systems disabled. The ATP system was still under trial and might have been expected to be disabled. The inquiry focused its

attention on the absence of the AWS, and the fact that the train driver was left to nego-
tiate one of the busiest routes in the country at high speed relying purely on visual
recognition of signals to which a correct response had to be made every time. In spite
of the shortcomings of the AWS, it will usually draw the attention of the driver to the
signal aspect ahead. The probability of passing signals with a cautionary or danger
aspect must be greatly increased without the presence of the AWS system, especially
when drivers are used to relying on it. Although there is no direct evidence, it seems
quite possible, that when drivers are preoccupied with their tasks that they might come
to rely on the AWS to warn them when a cautionary or danger signal is being
approached. In spite of the ruling that AWS is only a driver's aid, once this dependence
is established, it seems likely that driving without the system will make a SPAD more
probable. It is certainly possible that if the AWS had been working on that day, that the
accident would not have occurred. The decision to allow this train to operate without
the AWS was clearly a management decision for which they must take responsibility.

10.5.3 The Ladbroke Grove accident 1999

10.5.3.1 *The accident*

Only a few weeks after the inquiry into the Southall accident had opened, a similar accident took place on 5 October 1999 at Ladbroke Grove. It resulted in 30 fatalities, and
was the most serious rail accident for a decade. Like Southall, it occurred on the busy
approaches to Paddington station and involved a HST of Great Western Trains travelling
from Cheltenham Spa to Paddington. It was again caused by a driver passing a signal at
danger, but this time it was not the HST that was at fault but a three-car Thames Trains
local service travelling from Paddington which passed through a red signal SN109 into
the path of the express. Following the collision, fires broke out due to the ignition of
leaking diesel fuel. As a result 31 people died, 24 from the local train and 7 from the
HST, including both drivers, and a further 227 passengers were taken to hospital.

Immediately prior to the accident, a railway control room operator monitoring the
position of trains in and out of Paddington had realized that the local train had passed
SN109 at red and had run on to the section of the track along which the HST was travelling in the opposite direction. He therefore set the signal into Paddington ahead of
the HST to red to warn the incoming driver. Another operator warned the HST driver
of the train in his path using cab radio. However by the time these warnings were
received the HST was at the red signal, and there was insufficient time to bring it to a
halt. The combined collision velocity was estimated to be 135 m.p.h.

10.5.3.2 *The inquiry*

The accident at Ladbroke Grove took place a few weeks after the Public Inquiry into
the Southall accident opened. The Health and Safety Commission immediately ordered
an investigation and announced that a Public Inquiry into the accident would take
place under Lord Justice Cullen. The Inquiry Report (Cullen, 2000) examines a number
of issues, including the crashworthiness of rolling stock, the incidence of fires and the

means of passenger egress from trains in an emergency. As in the Southall accident, the HST had been fitted with a prototype ATP system but this was not operative at the time. However, it was concluded that its presence would not have had a significant influence upon the course of the accident since it was the local train which had passed through a red signal. The local train was fitted with the normal AWS system and as far as could be ascertained from the wreckage this was active and no faults could be found. The actions of the driver of the Thames Trains service from Paddington were examined in some detail at the inquiry.

It was found that the driver of the local train had cancelled the AWS alarms at the single yellow and double yellow aspect signals preceding the red signal SN109. Data recovered from the train indicated that just before reaching SN109 the driver had increased power to the train immediately after cancelling the AWS warning. He then passed the signal at about 40 m.p.h., and accelerated to 50 m.p.h. until he was less than 100 metres from the approaching HST. Emergency braking was then applied but the collision could not be avoided.

A thorough investigation was made into the condition of the signalling system in the area. Although no operational faults were found, there were considerable difficulties in the sighting of signals by drivers due to an unsatisfactory layout. Signal SN 109 was already on a Railtrack list of 22 signals which had been subject to frequent SPADs. Eight SPADs had occurred at that particular signal between August 1993 and the date of the accident.

The inquiry also found that the driver of the local train was relatively inexperienced, having been recruited in 1999. He had only recently passed through the driver training programme operated by Thames Trains. This comprised a 25-week period of initial training followed by 6 weeks of specific route learning accompanied by an instructor. He had been assessed as a competent driver on 15 September 1999 for that particular route and could therefore be regarded as a novice. He had only completed nine shifts on the route as the driver in charge prior to the accident. The Inquiry Report states that there is no evidence that the driver was fatigued or that there was 'a decline in alertness due to monotony, given that the train passed signal SN109 only 3 minutes after leaving Paddington'. At first sight it seems there is no explanation for the cause of driver's error. It is only when the difficulties of reading signals in the Paddington area are examined that the driver error becomes more understandable.

Following the Southall and Ladbroke Grove enquiries there were many conflicting views about the relative merits and costs of various train protection systems which could prevent such accidents again in the future. In order to resolve these uncertainties a special Commission was set up to investigate aspects of train protection and make recommendations (Uff and Cullen, 2001). One of the conclusions of the Commission concerning the Ladbroke Grove accident was that 'the complexity of the layout and the signal gantries, the range of approaches, and the obscuration of the signal aspects by all the line equipment presented an exceptionally difficult signal reading task'.

The crucial signal SN109, along with many other signals in the area, was mounted on a gantry which was obscured on the approach by girders below the deck of a road bridge. The red aspect of signal SN109 was only visible 60 metres beyond the point where the other signals on the same gantry became visible. To make matters worse, at the time of the accident the sun was shining directly on SN109 and this reduced its contrast compared with the surroundings so it was even less likely to be observed. The result was that the sighting time of the signal at normal line speed was between 5 and 7 seconds, this being the time window in which the driver must see the signal to make a response. Five to seven seconds is a very short period when it is also considered that this novice driver also needed to glance downwards at the cab controls from time to time to monitor the speed of the train. In these circumstances, a failure to notice the red signal does not seem so unusual.

This is typically one of the problems which the driver's AWS is intended to overcome. When the driver is distracted, or for some other reason fails to notice a cautionary or danger signal, the AWS reminds him that a signal is being approached. Yet this driver had already cancelled the warnings for the preceding single yellow and double yellow signals but had not taken any action to reduce speed. In fact, after passing the single yellow signal the driver accelerated towards the red signal. There is no official explanation why he cancelled the AWS alarm triggered by the final danger signal. A possible explanation is that the final failure to respond to the AWS was a dependent error as described in Section 10.4.3.

10.5.3.3 *Conclusion*

The direct cause of the accident was the failure of the driver of the local train to respond to a red signal, after having cancelled the AWS for that signal and the preceding cautionary signals. The root causes can be identified in the inadequate systems which failed to provide the driver with proper support for his task, these being the clear responsibility of management. The accident was due to a combination of the following factors:

(a) a novice driver who had only driven nine shifts on this route,
(b) inadequate signal layout with obscuration by bridge girders,
(c) minimal time to observe the red signal aspect,
(d) sun shining on the red signal light diminishing its relative intensity and contrast,
(e) the shortcomings of the AWS driver aid, in particular the increased probability of a dependent SPAD error following cancellation of the AWS at the preceding cautionary signals.

When all these factors are taken into account, then the occurrence of this SPAD is revealed as a systemic error whose probability was increased by an element of dependency due to the AWS system.

10.6 Human error analysis of signals passed at danger

10.6.1 Introduction

This section presents a quantitative human error analyses of SPADs. It calculates the overall probability that a train driver will commit a SPAD based on recent annual statistics. However, before a driver passes through a red signal he will also have passed through a series of cautionary signals. Thus, the overall error probability is in reality the joint probability of a number of separate errors. The probability of the separate errors is estimated by developing a human reliability analysis (HRA) event tree (as described in Chapter 6) for the whole sequence of encountering cautionary and danger signals. This section also provides, therefore, a practical illustration of the use of HRA event trees to model an action sequence and increase the understanding of the various interactions.

10.6.2 Causes of SPADs

The causes of SPADs have been analysed using a classification of error types devised by HMRI. These are broken down into categories of misreading, misjudgment, disregard and miscommunication. Of interest for this section is the category 'disregard'. Figures produced by Railtrack (Ford, 2001) show that 58 per cent of SPADs where the driver is responsible fall into the category of 'disregard' error, attributable to:

- failure to react to a caution signal (24 per cent),
- failure to check signal aspect (16 per cent),
- failure to locate signal (7 per cent),
- anticipating signal clearance (5 per cent),
- violations (6 per cent).

It is interesting that nearly half the 'disregard' errors are caused by failure to react to a caution signal. Such errors fall into the category of dependent failures, where an error is made more likely because of the incidence of a previous error (see Section 10.4.3). The coupling mechanism between the error of passing through a caution signal and later through a red signal, may be due to the AWS cancelling action being a conditioned response, distraction or work overload.

10.6.3 Probability of SPADs

Just prior to the Ladbroke Grove accident, HMRI revealed that in the previous year there were around 600 SPAD incidents of varying severity, representing an average of nearly two per day and an 8 per cent increase on the previous year. After the Southall and Ladbroke Grove accidents it should have been clear that a serious problem existed which could not be solved simply by disciplining or prosecuting the driver. Attribution of blame ignores the fact that accidents are being caused by systemic errors. The attempted prosecution of the driver from the Southall crash shows this was still not fully recognized.

From Railtrack and HMRI statistics, it is possible to estimate the probability of a SPAD. The probability of a human error can be calculated, as described in Chapter 4, by dividing the number of errors which occur by the number of opportunities for error. It is estimated that about 11,500 UK train drivers encounter about 10 million red signals per year. Thus the opportunities for error experienced on average by a driver are about

10,000,000/11,500 = 870 red signals per year. In the year ending March 2001 there were 595 SPADs, which, averaged over 11,500 drivers, approximates to about 0.052 SPADs per driver per year. This approximates to 1 SPAD for every 19 years that a driver is employed.

The average probability of a SPAD can therefore be calculated by dividing the number of errors (0.052 SPADs per year per driver) by the number of opportunities for error (870 red signals encountered per year per driver). The driver error probability is then:

$$P_{\text{SPAD}} = 0.052/870 = 5.95E-05 \text{ or approximately } 6.0E-05$$

This amounts to about one SPAD per 17,000 red signals encountered. Of course, the same value can be obtained by dividing the total number of red signals in a year into the annual number of SPADs. It has been calculated on a 'per driver' basis to indicate the number and frequency of SPADs an average driver may commit. This, of course, is just an average and some drivers will experience many more SPADs than others. The number of SPADs experienced by a driver is just as likely to be caused by the route as by the driver, since some routes will be busier than others or have a higher proportion of signals which are difficult to observe. Also, the problem of SPADs will be worse for HSTs than for local trains. A HST travelling at 125 m.p.h. will, for a four aspect signal system, with signals set at 0.75 mile intervals apart, pass a signal every 21.5 seconds. The braking distance of a HST is about 1.25 miles. This allows little margin for error on the part of the driver.

10.6.4 Human reliability event tree for SPADs

10.6.4.1 *Introduction*

Figure 10.1 represents a notional HRA event tree for a four aspect colour light signalling system. The labelling convention and method of construction and calculation is as described in Chapter 6, Section 6.2.2. The total probability of a SPAD based on statistics was shown above to be 6.0E−05. However this is a joint probability derived from the probabilities of the various failure paths that can lead to a SPAD, including dependency effects. Each failure path represents a possible scenario whereby a SPAD can occur. The purpose of the HRA event tree is to show the failure paths of each scenario and to assign notional probabilities to the various errors that lead to a SPAD. The notional probability data used to solve the tree is shown in Table 10.1. The three main scenarios or failure paths represented lead to outcomes F_1, F_2 and F_3 and their derivation is described below.

10.6.4.2 *Scenario 1 – Leading to failure path outcome F_1*

This scenario represents the outcome where the driver passes and ignores a double yellow signal (error A) followed by single yellow (error B) and red (error C) signal in turn. It is assumed that events B and C are dependent errors as described in Section 10.4.3. Error B is assumed to have a medium dependency (MD) on error A having

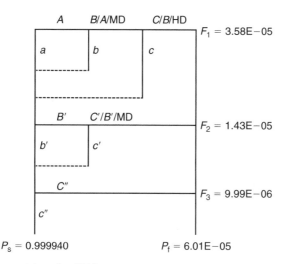

Figure 10.1 HRA event tree for SPAD.

occurred, and is thus shown as B/A/MD. Once error A and error B have occurred the level of dependency is assumed to increase from medium to high. Error C therefore has a high dependency (HD) on error B having occurred and is shown as C/B/HD. The same notional basic human error probability (BHEP) value has been used for A, B and C and assigned a value of 5.0E−04, the median of the range of error probabilities for a skill based task, described in Section 2.3.1.1.2 in Chapter 2.

The error probabilities used in the event tree are shown in Table 10.1, with the values of error B and C being increased due to dependency using the technique for human error rate prediction (THERP) dependency model described in Chapter 5. The overall probability of outcome F_1 in Figure 10.1 is calculated to be 3.58E−05.

10.6.4.3 *Scenario 2 – Leading to failure path outcome* F_2

This scenario represents the outcome where the driver observes and acknowledges the AWS alarm for the double yellow signal (success event a), but ignores the single yellow (error B') and Red (error C') signals which follow. It is assumed that events B and C are dependent errors as described in Section 10.4.3. Error C' is assumed to have a MD on error B having occurred, and is thus shown as C'/B'/MD. The probability values assigned for errors B' and C' will differ from the values for errors B and C used in Scenario 1 above. This is because if a signal is observed and acknowledged, then the probability of ignoring the following signals is likely to be less. This is, in effect, a 'reverse' dependency effect, sometimes referred to as 'negative dependency'. The dependency effect discussed above where a probability is increased because of a previous error, would then be properly referred to as 'positive dependency'. The notional BHEP value used for errors B' and C' has been assigned the reduced value of 1.0E−04 as shown in Table 10.1. The value of error C' is increased due to (positive) dependency using the

Table 10.1 Data for SPAD HRA event tree

Event reference number	Success/ Failure	Description of error	Probability P_s/P_f	
			Basic HEP	HEP with dependency
1	a	Driver ignores double yellow	0.9995	0.9995
	A		5.00E−04	5.00E−04
2	b	Driver ignores single yellow	0.9995	0.8567
	B		5.00E−04	0.1433
3	c	Driver passes red signal	0.9995	0.4998
	C		5.00E−04	0.5003
4	b'	Driver ignores single yellow (given a)	0.9999	0.9999
	B'		1.00E−04	1.00E−04
5	c'	Driver passes red signal (given a and B')	0.9999	0.8571
	C'		1.00E−04	0.1429
6	c''	Driver passes red signal (given a and b')	0.99999	0.99999
	C''		1.00E−05	1.00E−05

THERP dependency model. The overall probability outcome F_2 in Figure 10.1 is calculated to be 1.43E−05 based on the above assumptions.

10.6.4.4 *Scenario 3 – Leading to failure path outcome* F_3

This scenario represents the outcome where the driver observes and acknowledges the AWS alarm for the double yellow signal (success event a) and single yellow signal (success event b) but ignores the red (error C'') signal which follows. Due to the success in observing both of the caution signals, a lower probability is assigned to error C'' than to error C' in Scenario 2 due to negative dependency. A value of 1.0E−05 has therefore been assigned, compared with 5.0E−04 for error C'. Obviously there is no positive dependency since no previous errors have occurred. The overall probability outcome F_3 in Figure 10.1 is calculated to be 9.99E−06 based on the above assumptions.

10.6.4.5 *Total failure probability* P_f

A SPAD can occur as result of Scenario 1 or Scenario 2 or Scenario 3. Hence the total probability of a SPAD, P_f is calculated from the sum of outcomes F_1, F_2 and F_3 using the addition rule of probability theory (see Section 5.2.3). The product term $(F_1 \times F_2 \times F_3)$ can be ignored because F_1, F_2 and F_3 are very small:

$$P_f = F_1 + F_2 + F_3 = 3.58\text{E}-05 + 1.43\text{E}-05 + 9.99\text{E}-06 = 6.01\text{E}-05$$

The event tree shows that the greatest contribution to the overall SPAD probability comes from Scenario 1. In Scenario 1, once a failure to observe a double yellow signal (assumed probability of 5.0E−04) has occurred, the probability of failing to observe the single yellow signal becomes 0.1433 with dependency, or about 1 in 7. As there have now been two failures to observe cautionary signals, the dependency is increased still further so that the probability of failing to observe a red signal becomes 0.5 or about 1 in 2. The contribution of Scenario 1 is therefore about 60 per cent.

Scenario 2 contributes about 24 per cent of the overall probability. This is smaller because the probability of failing to observe a single yellow signal (after having observed a double yellow) has been reduced to $1.0E-04$, about one fifth of the previous value due to the effects of negative dependency as described above. However, because of this failure the effect of positive dependency then increases the probability of failing to observe a red signal from $1.0E-04$ to 0.1429, again about 1 in 7. Scenario 3 therefore contributes the remainder of the overall probability equivalent to about 16 per cent.

Observant readers may note that not all possible scenarios have been included in the event tree. For instance there is a potential scenario where a driver may fail to observe a double yellow signal (error A), then observe a single yellow signal (success event b) before failing to observe the red signal (error C). However, this scenario is considered to be much less likely, would make only a small contribution to the overall probability and has therefore been omitted in the interests of simplicity.

10.6.5 Conclusion

Clearly the notional values of the separate probabilities for failing to observe double yellow, single yellow and red signals as given in Table 10.1 and used in the event tree for each of the three scenarios, have been chosen carefully. The intention was to use an HRA event tree to estimate what these separate probabilities might be in order to produce a total probability of a SPAD of about $6.0E-05$, equivalent to the probability calculated from SPAD statistics in Section 10.6.3.

For an initial failure to observe the first cautionary signal in a series of signals, an arbitrary probability of $5.0E-04$ was chosen, equivalent to about 1 in 2000 cautionary signals encountered. The assumed value of $5.0E-04$ is the median of the range of probability values given for skill based tasks ($5.0E-03$ to $5.0E-05$) in Chapter 2. It is skill based because the task of observing and responding to cautionary or danger signals is a simple task well practised by drivers in their everyday duties. However, the task also benefits from the AWS device to help maintain driver vigilance, and on this basis *if it was effective*, the probability might be expected to be at the lower end of the skill-based range. However, its effectiveness is already in question as discussed previously

It should be noted that the statistically derived overall human error probability (HEP) of $6.0E-05$ is averaged across all signals, irrespective of line speeds, visibility and sighting distance. The error probability of a SPAD at the twenty-two signals on the Railtrack danger list, for instance, will be much greater than this.

The HRA event tree in Figure 10.1 is believed to be a realistic model of the way in which SPADs occur in reality. It also provides a practical illustration of the power of an HRA event tree to analyse the errors in a sequence of tasks and increase the level of understanding of the various dependencies and recovery paths.

10.7 Driver protection against SPADs

10.7.1 Introduction

The accidents at Southall and Ladbroke Grove resulted in more attention being paid to performance shaping factors such as improvements to signal layout to give better sighting, driver training and a general raising of awareness among drivers of their susceptibility to SPADs. However, it will prove difficult significantly to reduce the probability of SPADs occurring by further improvements to the driver's task given the existing shortcomings of the existing AWS. It has been stressed earlier in this book that if an active human error can cause a catastrophic accident, with a high probability that the error cannot be recovered, then it is necessary either to eliminate the human from the system or introduce some reliable method of protection against the consequences of the error. In the case of SPADs one approach is to automate the task and take the human being out of the situation completely. Ultimately, this would mean the use of automatic train operation (ATO) based on ATP (see Appendix and below). However, there are a number of intermediate options that do not require the use of full ATO and these are described below.

10.7.1.1 *Double-block working*

One approach to the problem of reducing the probability of a collision following a SPAD is the use of the double-block system. The approach was investigated by HSE in their report 'Management of Multi-SPAD Signals' (Health and Safety Executive, 2000), from which the following is an extract:

Among the available measures to reduce SPAD risk, double-block working, which only allows the previous signal to be cleared when the signal in question has already been cleared, has the superficial attraction that it almost completely eliminates the possibility of SPAD, since the signal can only be encountered at danger if the previous signal has already been passed at danger. However double-block working can only be achieved at the expense of reduced line capacity which may result in more signals being encountered at red and may transfer increased risk to those other signals. Therefore, while it provides a useful yardstick against which to test other mitigating measures, the circumstances in which double blocking is applicable are very limited.

Double-block working effectively means that there needs to be two consecutive SPADs for an overrun to occur into a potential collision zone. The disadvantage of this approach is reduced line capacity and an increase in the number of red signals encountered which may therefore increase the number of SPADs. Another disadvantage, not mentioned in the report, is the dependency effect such that if the driver has already passed through a series of cautionary signals, and cancelled the AWS alarm, and then passed through the first danger signal, the probability that he will pass through the second danger signal is considerably increased. It is therefore unlikely that double-block working would significantly reduce SPAD accidents.

10.7.2 Train protection and warning system

The train protection and warning system (TPWS) (see Appendix, Section A.2.5) was devised as a an expedient means of providing a form of train protection against SPADS without incurring the major cost or disruption to rail services associated with ATP (see below). An advantage of TPWS is that it uses existing AWS equipment reducing not only the cost of installation, but also the time required to install the system across the network. In addition it is easier to retrofit to existing rolling stock and infrastructure. The system has two main features:

- a speed trap installed in advance of a red signal,
- automatic train stop at a red signal.

If a train is approaching a red signal at a speed which is estimated to be too high to prevent an overrun into the next section, the brakes are automatically applied, overriding any actions of the driver and preventing the occurrence of a SPAD. Even with TPWS installed, the AWS still provides the train driver with an audible and visual warning of the approach to a cautionary or red signal so that he can still slow down or stop the train. However, TPWS automatically applies the brakes of a train travelling too fast to stop at a red signal fitted with TPWS equipment and will operate regardless of whether the driver has cancelled the AWS warnings. It is therefore an independent safeguard against driver error.

At train speeds up to about 75 m.p.h., TPWS will bring the train to a halt within the overlap, that is, the distance beyond the signal where there is no danger of collision. Above 75 m.p.h., the possibility of a collision cannot be ruled out, but at least the system will ensure that a speed reduction is made, which will mitigate the consequences of any collision that might occur. TPWS is currently being fitted across the UK rail network, initially at signals where a significant SPAD problem has already been identified. Ultimately, the system will be fitted across the entire network until a more effective ATP system meeting European standards, as described below, can be installed to protect against SPADs at all train speeds.

10.7.3 Automatic train protection (ATP)

While the AWS system may well have proved effective as a driver's reminder appliance and prevented many SPADs occurring , its undoubted weaknesses have still left train drivers vulnerable to human error. The three major accidents described above (Purley, Southall and Ladbroke Grove) have highlighted these shortcomings and increased awareness of the problem to the extent that further action was demanded by government, regulators and the general public. As long ago as 1989, the Court of Inquiry into the Clapham Junction accident (see Section 10.2.2) recommended that a system of ATP be installed across the UK rail network to protect passengers from the hazards of signal overruns (although ATP would not have prevented the Clapham Junction accident). Implementation was intended to follow testing of the system on Chiltern Lines and Great Western Railway (GWR).

The end result was that the huge costs of installing ATP across the network was judged by the British Railways Board to be uneconomic in view of the relatively small number of passenger lives which would be saved. Hence in 1995, a decision was made not to install ATP. There was some justification for this decision. It was shown that a modest investment in central locking of carriage door on main line passenger trains could reduce the incidence of passengers falling to their death from trains in motion to almost zero. This cause of death amounted to half of all passenger fatalities across the rail network. The investment was made bringing about the greatest single benefit ever achieved in rail safety at minimal cost and disruption to services. By comparison, the cost of ATP in relation to the number of lives saved was disproportionate. The ATP concept was therefore effectively shelved until the issue of SPADs was raised again following the Southall and Ladbroke Grove accidents.

10.7.4 Conclusion

The main objection to the installation of ATP has always been the excessive cost per life saved and this has virtually ruled it out for the past decade. Nevertheless, as signalling systems are modernized across the network it may be more economic for ATP to be installed. One problem is the difficulty of retrofitting existing rolling stock with ATP. However, it is disappointing that even rolling stock currently under order or being supplied to train operating companies does not have any built-in ATP compatibility. This is hardly due to a lack of foresight, since Anthony Hidden QC recommended that ATP be installed nationwide after the Clapham Junction accident in 1989. Yet more than a decade has elapsed since the Hidden Report was published, a period which has seen numerous SPAD accidents up to and including the Ladbroke Grove accident in 1999. This astonishing state of affairs is principally due to a lack of will to make the necessary investment on the part of the railway industry using arguments of the high cost of ATP per fatality averted but exacerbated by a lack of a consistent policy by successive governments.

There are still many conflicting views about the relative costs and benefits of various train protection systems. A special commission was set up in 2000 to investigate all aspects of train protection (Uff and Cullen, 2001) in order to resolve the many issues. One of the issues was that ATP, as originally piloted by British Rail, is now old technology and has proved unreliable during the GWR trials. As a result, the ATP on trial on both HSTs involved in the Southall and Ladbroke Grove accidents had been disabled. Technologically, British ATP has been overtaken by a European system known as European Rail Traffic Management System (ERTMS) that a European interoperability directive requires must be fitted eventually on all UK main lines.

The Government has now agreed that railways must install ERTMS when highspeed lines are upgraded. In the UK it is probable that ERTMS will be installed on the Great Western Main Line, the West Coast Main Line and the East Coast Main Line, but this will take some time. Currently, the only railway line in the UK where a European compatible ATP system will be installed is the new Channel Tunnel Rail Link from London to Folkestone. There have been strong objections from the rail industry which maintains that the current version of ERTMS would cause significant disruption and reductions in services by about one train in eight. It is argued that this might result in many passengers using a car instead of the train leading to an overall increase in fatalities. The rail industry has therefore proposed delaying installation of ERTMS beyond 2010 allowing development of a system that will increase rail capacity and prevent disruption. This would result in the installation of ERTMS on high-speed lines but not earlier than 2015 and on all other lines by around 2030. The railway industry has also maintained that it would not be possible to pay for the upgrades from its own resources requiring substantial funding by the taxpayer and/or railway users.

In the meantime, the TPWS has been developed and tested and is gradually being installed across the network. It must be remembered, however, that TPWS would not have prevented the Southall accident, although it would probably have stopped the local train before it overran a red signal at Ladbroke Grove. The installation of TPWS is already having a beneficial effect on the incidence of SPADS. However, as the accident statistics in Chapter 7, Section 7.4.2, show the severity of accidents has increased in recent years due at least partly to higher train speeds. TPWS will not be so effective in reducing the incidence of high-speed accidents due to SPADs. Thus, the problem of SPAD errors in not yet resolved. If more high profile accidents occur the pressure to bring forward the installation of ERTMS or its equivalent will become irresistible.

References

Anthony Hidden QC, (1989). *Investigation into the Clapham Junction Railway Accident*, London: Department of Transport (HMSO).

Ford, R. (2001). The Forgotten Driver, *Modern Railways*, January 2001, pp. 45–48.

Health and Safety Executive, (2000) www.hse.gov.uk/railway/spad/manmss.pdf.

John Uff (2000). *The Public Inquiry into the Southall Rail Accident*, Sudbury, Suffolk: HSE Books.

Uff and Cullen (2001). *The Joint Inquiry into Train Protection Systems*, Sudbury, Suffolk: HSE Books.

Lord Justice Cullen (2000). *The Train Collision at Ladbroke Grove, 5 October 1999*, Sudbury, Suffolk: HSE Books.

11

Active errors in aviation

11.1 Introduction

Aviation is not very tolerant of human error. The margins of safety are narrow and small variations in human performance can be catastrophic. The effects of latent and dependent errors during aircraft maintenance have already been examined in Chapter 9. This chapter describes active errors that have led to accidents involving commercial aircraft. Most aircraft accidents take place during take-off or landing, at a time when there is the least margin for error. For most of the time an aircraft is in flight it operates on automatic pilot. The three main flight parameters of altitude, air speed and magnetic heading are continuously monitored and controlled so that the scope for pilot error is considerably reduced. On more modern aircraft, the automatic pilot itself is directed by a computerized Flight Management System (FMS) to follow a programmed route. The FMS is capable of flying the aircraft to the destination airport and lining it up with the runway ready for landing. The pilot will only take manual control of the aircraft to land it when it is about 50 feet above the runway. His only remaining task is to close the throttles and apply reverse thrust and braking.

As a result of improvements to aircraft reliability, coupled with computerized FMS and improved navigational aids, flying is by far the safest form of transportation based on risk of fatality per kilometre travelled, followed closely by railways. The passenger is more at risk when driving by car to the departure airport or station when the risk of fatality per kilometre is about 100 times greater. When an air accident does occur, however, it usually involves multiple fatalities and there is much greater public aversion than if the same numbers of fatalities had occurred singly on the roads. Risk aversion generally means that the acceptable expenditure to reduce the risk of multiple fatality accidents is much greater per fatality than the expenditure which will be necessary to prevent the same number of fatalities occurring separately.

The case studies in this chapter examine the disastrous effect of some deceptively simple active errors. The first case study describes the consequences of an error in setting a rotary switch in the wrong position and not noticing even though a check on other aircraft parameters should have immediately revealed this. The result was the loss of a Boeing 747 (B747) aircraft over Arctic waters, the deaths of 269 passengers and crew, and an international incident. The second case study involved the problem of laterality or handedness, where a pilot failed to distinguish between left and right due to a poor instrument layout in the cockpit. This accident was the last major aircraft crash to occur in the UK and led to the deaths of 39 passengers. The difference between an error and an accident is usually the absence or failure of recovery. Hence, in order to reduce the probability of an accident not only is it necessary to identify improvements to reduce the probability of the active error, it is also necessary to identify the reasons why recovery did not take place. Thus, in these case studies, as much attention is given to recovery as to why the error occurred in the first place.

11.2 The loss of flight KAL007

11.2.1 Introduction

The accident described below illustrates how a simple unrecovered active error can lead to catastrophe. It also resulted in a major international and diplomatic incident which made headlines across the world and resulted in heightened tension between the USA and USSR interrupting the nuclear disarmament process and prolonging the Cold War. At first sight, the active error may appear to be latent, since the consequence was delayed by some 5 hours. However, the effects of the error were immediate, resulting in an aircraft gradually diverging from its assigned route. The divergence increased until recovery was no longer possible and disaster could not be avoided. The error therefore falls into the category of an active and not a latent error.

11.2.2 The accident

Korean Airlines flight KAL007, a B747 passenger jet, took off from Anchorage, Alaska on 31 August 1983 at 03.30 hours local time bound for Seoul, South Korea with 246 passengers and 23 crew on board. It was before the end of the Cold War and the route the aircraft took was within 400 miles of forbidden territory east of the Soviet Union. The shortest route from North America to the Eastern cities of Tokyo or Seoul follows a 'great circle', this representing the shortest geographical distance between two points on the globe. In this case the great circle skirted the edge of the vast wasteland of the Arctic. Unnoticed by aircrew and ground controllers, as the flight progressed it gradually drifted off course so that by 4 hours into the flight it had strayed 365 miles west of its expected position and into prohibited Soviet territory in the vicinity of the Kamchatka Peninsula.

At the same time as the KAL007 flight was en-route, a US Boeing RC-135 intelligence plane was also flying in the same airspace on the edge of Soviet territory, and was being closely monitored by Soviet military radar. As flight KAL007 approached the Kamchatka Peninsula, Soviet defence forces mistook the radar echo of the passenger jet for the RC-135 and as a result, six MiG-23 fighters were sent to intercept. At this point KAL007 left Russian airspace over the Okhostk Sea and the fighters were ordered to return to their base. KAL007 was now headed for the Soviet territory of Sakhalin Island, with the pilots still unaware of their true position. Once more, Soviet fighter aircraft were scrambled and two Sukhoi Su-15 fighters from the Dolinsk-Sokol airbase were sent to investigate.

As KAL007 again entered Soviet airspace, the fighters were ordered to attack. At 18.25 hours GMT, in the darkness over Sakhalin Island, and just before it left Soviet territory, the B747 was brought down by two heat-seeking missiles from a Soviet jet, piloted by Major Gennadie Osipovich. The B747 aircraft immediately lost cabin pressure and as a result of control problems, spiralled into the Arctic waters below with the loss of all on board. The flight crews of aircraft operating in this vicinity were fully aware that there was a preponderance of Soviet military facilities in the vicinity of Kamchatka Peninsula and Sakhalin Island and if they entered prohibited USSR airspace they could be fired upon without warning. The shooting down of KAL007 led to an international incident with the Soviet Union being accused by the USA of acting in a reckless manner. This was met by counter-accusations claiming that the aircraft was on a spying mission.

11.2.3 The causes

No official explanation was ever provided by the US authorities about why the flight went so far off course. As a result, many unofficial theories were put forward to explain why the disaster took place. These included not only technical explanations, but also numerous conspiracy theories suggesting that the aircraft was either lured into Soviet airspace or had purposely invaded it for nefarious purposes which will not be explored here. These theories resulted in the publication of numerous books and papers. Up until recently the Russian Federation still insisted that the aircraft was involved in a spying mission.

There is however, a very simple explanation – the aircraft was off course due to a single human error which took place at the start of the flight. The explanation given below is based on the findings of independent investigations carried out by the International Civil Aviation Organization (ICAO, 1993) in a report published some 10 years after the disaster.

Most modern airliners are equipped with four independent means of navigation, any of which can give the position of the aircraft to standards of accuracy varying from a few miles to a few metres. The investigations centred upon how it was possible for the aircraft to be so far off course without the knowledge of either the flight crew or air-traffic controllers given these modern navigation aids. The direct cause of the disaster was clearly the shooting down of the aircraft by Soviet fighters (something that was almost inevitable in view of the aircraft's position). The root cause was to be found in a series of errors concerning the operation of the aircraft's navigational systems. In order to understand these errors it is necessary to examine the navigational aids available to the pilot.

11.2.3.1 *Navigational systems*
11.2.3.1.1 The magnetic compass
Before about 1970, the principal navigation system used by civil aircraft was the magnetic compass, backed up by sightings of the sun or stars; more or less the same technology used by Christopher Columbus to reach America. The magnetic compass has a number of advantages and disadvantages. Its advantages are that it is reliable in operation, it has instant availability and it does not require a power supply in order to operate. Due to this, it is carried as a mandatory item by all aircraft because if all else fails, the magnetic compass can still be used. Its chief disadvantage is that it is only accurate over very short journeys because of the effect of wind drift. It is also useless near the magnetic poles. Thus on a long flight, the pilot must take a manual fix on the sun or stars from time to time using a sextant in order to correct any deviation from the intended course. This effectively provides a built-in check, because if an error is made in any one fix, the next fix or two will quickly reveal the mistake.

11.2.3.1.2 The Non-Directional Radio Beacon
Another older system of navigation used on aircraft is the Non-Directional Radio Beacon (NDB). This comprises a simple radio transmitter set on a hilltop sending out a call sign. When within range it is detected by a radio compass on the aircraft. The compass indicates the direction of the beacon and the aircraft is able to fly towards it. It has the advantage of simplicity, but over long distances it is not very accurate.

11.2.3.1.3 Very High Frequency Omni-directional Radio Beacon
A more modern form of radio beacon is the Very High Frequency Omni-directional Radio (VOR) Beacon. This comprises a multidirectional radio antenna which transmits pencil thin beams of radio waves called 'radials', at a known frequency and from a known position. An aircraft fitted with VOR equipment is able to lock on to the radio beam, and is then said to have 'captured a radial'. The VOR instrument on the aircraft indicates the direction to the VOR Beacon. Thus, unlike the NDB, it is not necessary for the aircraft to fly direct to the beacon, but by tracking across the radials, it is possible for the aircraft to follow accurately a known course in relation to the beacon. VORs are often fitted with distance measuring equipment (DME) which indicates the distance of the aircraft from the VOR Beacon. Two VOR instruments are provided and if the aircraft is able to lock on to two radials simultaneously then this enables the position of the aircraft to be plotted to within an accuracy of 1 mile in 60 miles.

11.2.3.1.4 Inertial Navigational System
The principal method of navigation used on modern aircraft is the Inertial Navigational System (INS). Most navigation relies on relating the position of an aircraft at any time to a fixed point of reference outside the aircraft. For a compass, this fixed point is the Earth's magnetic pole. The INS uses an internal reference point, a tiny platform stabilized by gyroscopes which is unaffected by the earth's movement. Thus an INS can provide the pilot and the aircraft's FMS with information about the earth's rotation, the latitude and longitude of the aircraft, its height and whether it is flying a level and straight course. Larger civil aircraft usually carry three INS instruments, and operate on

a two-out-of-three voting system so that an error in one system is detected by the other two, and ignored for navigational purposes.

The combination of INS and VOR is therefore extremely accurate and simple to operate. This has meant that modern airliners no longer need to carry a third person in the cockpit undertaking the duties of a navigator. Navigation is carried out by the pilot and co-pilot as part of their normal duties. Before the advent of modern navigational systems when a navigator was employed in the cockpit it was not unusual for gross errors to be made, and it is estimated that about 1 in every 100 trans-ocean flights would end up as far as 30 miles off-track. However, as far as is known, an undetected failure of a two-out-of-three INS system has never yet occurred. The main errors in the use of INS systems have been in programming the co-ordinates at the beginning of the flight and a US study has identified the frequency of these errors at about 1 in 20,000 operations. Flight KAL007 possessed all the latest navigational equipment including INS and VOR.

11.2.3.2 *The flight*

The route taken by the aircraft is shown in Figure 11.1. The flight time from Anchorage to Seoul was estimated to be 8 hours 20 minutes. On take-off the aircraft was instructed to climb to 31,000 feet and 'proceed to Bethel, as able'. Bethel was a hamlet about 300 miles from Anchorage, and as well as a way-point was also a VOR Navigational Beacon. However, just west of Anchorage was Mount Susitna, and in order to miss this the pilot would fly a initial heading of 220 degrees until after crossing the 246 radial transmitting from the VOR at Anchorage Airport. Prior to take-off, the INS had been calibrated and was already programmed to fly to Bethel by the shortest route. According to the pilot's manual, when the aircraft had cleared the mountains, one of the two autopilots had to be selected to the 'INS mode setting' so that the aircraft would automatically be flown via the pre-programmed INS route. As the plane crosses the 246 radial, the actual heading is compared with the INS heading. If they agree then the INS has been set correctly. This system should be virtually foolproof if followed correctly. Unfortunately on the night of the 31 August, the VOR at Anchorage was shut down for maintenance and the VOR at Bethel was still out of range.

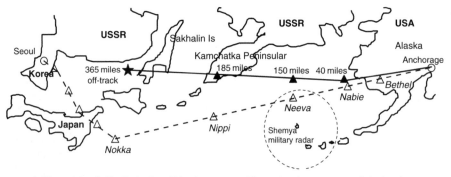

△ Way-points   ▲ Nautical miles off-track   – – – – Planned route   ———— Actual route

Figure 11.1 Route of KAL007.

Although it was not possible to check the INS setting in this way, there were a number of alternative options open to the pilot:

1. He could request a control tower check on the bearing of the radar image of the aircraft.
2. He could make a timed check on progress although this would be quite inaccurate.
3. He could use the INS to provide a bearing.
4. He could lock on to the NDB on Cairn Mountain, halfway to Bethel, and then capture the Bethel VOR radial and use this to check the INS.

This latter option was clearly the best, but in order to get within range of the Bethel VOR, the pilot would need to switch the autopilot to magnetic heading 246. However, as KAL007 passed by Bethel it was seen to be 12 miles off-track by the military radar at King Salmon, 175 miles south of Anchorage. The flight was obviously not flying on a radial by Bethel or by INS because of the gross error over this short distance.

To follow the later course of the flight, it can be seen from the map of the route, that in order to reach Seoul by the most direct route, the aircraft would need to fly on a great circle route (shown as a continuous line) following a constant compass heading of 246 degrees.

It can also be seen that such a heading would take the aircraft over the sensitive Soviet territory and the nuclear military installations on the Kamchatka Peninsular. Hence the co-ordinates (latitude and longitude) of a number of way-points needed to be programmed into the INS system to give a route which was longer but would avoid passing over Soviet territory. This route is shown as a dotted line and provides some clue to the extent of the course divergence that night.

When a route is programmed into the INS, then a system of reporting alerts is activated. One minute before arrival at each programmed way-point, the system generates an amber 'alert' light which warns the pilot that he will shortly need to make a position report to air-traffic control (ATC). A few minutes later, the light goes off, indicating that the way-point has been passed, and the pilot reports his position to ground control and gives his estimated time of arrival (ETA) at the next way-point. This system has one serious drawback, the alert light illuminates when the aircraft is abeam of the way-point even if the way-point is up to 200 miles away from it. As KAL007 passed the Bethel way-point, the amber alert light was illuminated and this was reported. The position was also reported at the following way-points of *NABIE*, *NEEVA* and *NIPPI* shown in Figure 11.1. At *NIPPI*, the pilot reported that he expected to reach the next way-point at *NOKKA* in 79 minutes. In fact it is now known that at way-point *NIPPI*, KAL007 was already 185 miles off-track, but just within the limit in which the INS alert light would still report him as being at the way-point.

11.2.3.3 *The active error*

There was a strong belief by the aircrew that the aircraft was still on course since position reports had been made at intervals along the route as far as *NIPPI*. To make position reports in this way without checking the actual position is quite acceptable so long as the INS system is in fact coupled to the autopilot, and the autopilot is actually steering the correct course. However, if the autopilot selector switch were to be wrongly selected to a different navigational mode, then the aircraft may in fact be flying on an entirely

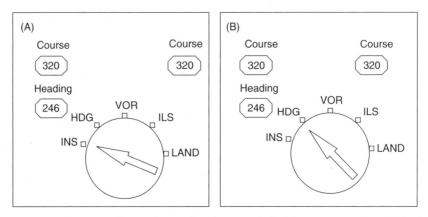

Figure 11.2 Schematic of autopilot mode selector switch: (A) Correct setting switched to INS and (B) Incorrect setting switched to HDG.

different course. The INS would continue to give an alert at each way-point if within the 200 miles limit. This is what happened with KAL007. The way this error is possible is shown in Figure 11.2.

The navigational mode of a B747 of this era (which was before the era of the glass cockpit, whose problems are examined in another case study later) was selected by means of a simple five position rotary switch on the instrument panel operated manually by the pilot. For the INS to be connected to the autopilot, the switch needs to be set to the 'INS' position, a simple and unambiguous task. Figure 11.2 shows that this switch needs to be displaced by only one notch upwards from the INS setting to be set to the HDG or magnetic compass heading. The autopilot is then linked to the magnetic compass. The other possible settings are VOR, where the autopilot will track the current VOR radial; ILS, where the autopilot is linked to the instrument landing system and LAND, where the autopilot is linked to the auto-land system.

Apart from visual observation of the position of the switch, there is no other indication of the mode that has been selected. Bethel lies on a magnetic compass heading of 246 degrees magnetic from Anchorage and this heading would need to be taken once the aircraft had flown around Mount Susitna. Allowing for winds, this was also the heading being followed when the plane was shot down. The ICAO investigation concluded therefore that:

- The resulting track deviation was due to the crew's failure to note that the autopilot had either been left in heading mode or had been switched to INS mode when the aircraft was beyond the range (7.5 nautical miles) for the INS to capture the desired track.
- The autopilot was not therefore being controlled by the INS as assumed by the flight crew.
- Manual control of the autopilot was not exercised by the flight crew by the use of heading selection (i.e. although it was possible manually to navigate the aircraft along the correct track using the autopilot coupled to the heading (HDG) mode, this was obviously not done).

● The maintenance of a constant magnetic heading was not due to any aircraft system malfunction.

The main conclusion of the ICAO report was that the accident was due to pilot error and 'the failure to detect the aircraft's deviation from its assigned track for over 5 hours indicated a lack of situational awareness and flight deck co-ordination on the part of the crew' (ICAO, 1993). The Shemya US military radar installations in the Aleutian Islands, which would normally be expected to have detected all aircraft in the vicinity, denied any knowledge of detecting a westbound aircraft crossing the Alaskan identification zones that night.

Such errors in the operation of the navigation system on an aircraft are not unknown. However, they would usually be detected by the flight crew before a significant deviation from the expected track had been made. In fact the operating manual provided by Korean Airlines allowed for such a possibility and called for multiple checks of the magnetic heading to be made, with necessary course corrections at each way-point. This was to ensure that the aircraft followed the track prescribed by the computer flight plan as well as ensuring that the pilots were observing current information from the aeroplane's navigation systems. In short, there were a number of independent mechanisms for detecting such an error and for recovery from it. The pilot could, for instance, have put his radar to 'mapping mode' to show outlines of landmasses below. For this flight, the presence of land would have meant prohibited Soviet territory. However, the company's operating manual did not require this check to be made.

Other aspects of the system conspired to make an error more likely. One of the more misleading system characteristics was the one mentioned above. The INS alert lights would purport to show the position of the aircraft at each way-point, even though the autopilot was selected to a different system and flying a different route, due to the 200 miles range within which it would still work. It is probable that the pilot's excessive confidence in the INS system did not encourage them to check their position by other means during the flight. It was unfortunate that when KAL007 was shot down, it was due to report to the next way-point at *NOKKA* 1 minute later, in spite of the fact that the aircraft was off course by 356 miles. As the aircraft was outside the 200 mile range for the alert system to operate, the alert light would not have come on. They would then have been required to contact Tokyo ATC who would have informed the pilot that he was off course. Tragically KAL007 was in touch with Tokyo at the same moment that a Soviet fighter was frantically trying to make radio and visual contact. Three minutes later the aircraft was shot down.

11.2.4 Conclusions

11.2.4.1 *Human factors analysis*

The active error of setting a manual rotary switch to the wrong position and failing to notice is not uncommon and is certainly not restricted to aircraft cockpits. A colloquial term for this type of human error is 'finger trouble'. The technique for human error rate prediction (THERP) methodology described in Chapter 4, Section 4.3.1.3 clearly identifies an error 'set a rotary control to an incorrect setting' and suggests a human error probability (HEP) of 0.001 or 1.0E−03 with an error factor of 10. This means that the chance of this error can vary from 1 in 100 to 1 in 10,000 with a mean value of 1 in 1000 operations. Using the mean value, if an aircraft made 500 flights per year, then this type of error would be expected to occur about once every 2 years. Multiply this by

Table 11.1 Qualitative human error analysis of the KAL007 accident

Cause	Description	Type of error or other failure	Systemic cause or preventive action
Direct cause	Aircraft was shot down by two Soviet missiles after inadvertently flying over sensitive Soviet military installations	Military pilots were operating under instructions from Soviet ground control	Failures within three countries can be identified: 1. Aircraft was not effectively tracked by the US military radar. 2. The Japanese Defence Agency were aware that an aircraft was tracking into USSR airspace but did not suspect it was a civil aircraft off its intended track. 3. USSR military did not comply with the ICAO recommended practices for interception of civil aircraft before attacking KAL007 (although confused by presence of the US RC-135 reconnaissance aircraft).
Root cause	Failure of the pilot to correctly set autopilot mode selector switch to INS setting	Active error	The mode selector position was not clearly marked and therefore not obvious to the pilot (faulty ergonomics and layout). The INS system was considered highly reliable (two-out-of-three voting system) but this assumed the INS was in fact connected to the autopilot. The system could be defeated by a single active error (over-confidence).
		Design error	The 200 mile limit within which the INS would continue to provide an alert that a way-point was being passed was misleading to the pilots. It would have provided an ideal error recovery mechanism but for this feature (system design fault).
	Failure to carry out routine position checks on INS as required by company operations manual	Violation	Management enforcement of procedures, audit and pilot check flights.
Contributory causes	Pilot fatigue	Loss of vigilance	The flight crew were physically fit but extended time zone crossings and the level of utilization of crew flight and duty times had the potential for one or more of the flight crew to experience fatigue and a reduction of situational awareness.

the number of aircraft in operation with similar controls then wrongly positioning a navigational mode selector must be an extremely frequent event. The fact that the error remained undetected over a period of 5 hours was the unusual feature of this accident. It demonstrates the importance of engineering recovery mechanisms into the aircraft systems and into flight procedures so that errors can be detected by means of routine checking. It was the fact that required routine checks were omitted on this flight, partly due to the violation of operating procedures by the pilots, and partly due to navigational radio beacons being unavailable, that this tragic accident happened at all. Table 11.1 presents a qualitative analysis of the causes and the main errors involved.

11.2.4.2 *Recommendations*

A number of recommendations were made by the ICAO report into the disaster:

- All aircraft following the Northern Pacific route from the USA to Central Asia (the second busiest route in the world after the North Atlantic) are now monitored by *civil* controllers manning the military radar at Shemya Island.
- A new radar point has been established at the entry point to the ocean route.
- In 1985 a trilateral Memorandum of Understanding was signed between Japan, the USSR and the USA. This involves the installation of dedicated voice communications between Anchorage, Khabarovsk and Tokyo, together with operating procedures which would avoid future misunderstanding.
- All aircraft should be fitted with a clear indication when the autopilot and flight director are in the HDG mode and in addition procedures for long-range navigation should include a specific check that the INS has captured the navigation mode before entering oceanic airspace.
- Airline operators should ensure that aircrew are fully aware of the consequences of leaving the autopilot flight director in HDG mode.

In spite of the fact that the flight crew were properly certificated and qualified for the flight and no fault could be found in the training procedures in use by Korean Airlines, 269 lives were lost as a result of an active human error. The reason that the error led to a consequence of this magnitude was the absence of checking tasks to detect the error and enable recovery to be made. As such checks were set out in operating procedures, the active error was therefore accompanied by a serious violation of rules. Whether this was due to pilot fatigue, or due to simple lack of attention and vigilance to the task in hand will never be known with absolute certainty.

11.3 The Kegworth accident

11.3.1 Introduction

The active error which caused this accident has some similarities to the one described in the previous case study with numerous missed opportunities for recovery. The case study is also an example of a 'mistake' or planning error (see Chapter 2). In a 'mistake', because of an early misdiagnosis, a plan of action is formulated which is inappropriate even though it may be carried out perfectly. Recovery was prevented by the occurrence of a 'mindset', a fixation on the initial diagnosis of the problem preventing it from being reviewed even when later evidence showed it to be wrong.

The case study also illustrates the importance of good design of the man–machine interface, in this case the cockpit display presentation, so that it fully supports the actions of the user. This is particularly important in emergency situations when there is an element of stress and high workload present. As noted in Part I of this book these are factors which can severely degrade performance. Into this mix of adverse performance shaping factors is also brought the issue of 'laterality'. This is an inability which affects everyone from time to time, but some more than others, to distinguish the left hand from the right hand, particularly under conditions of stress and when external stimuli are competing for limited attentional resources.

11.3.2 The accident

British Midland Airways flight BD092 departed from Heathrow Airport, London shortly before 20.00 hours on 8 January 1989 bound for Belfast. The aircraft was an almost new Boeing 737-400 (B737-400), a short-range twin-engine commercial jet airliner designed to carry up to 180 passengers. This aircraft had only received its Certificate of Airworthiness a few months previously, and at the time of the accident had flown for only 520 hours. The aircraft was installed with a 'glass' cockpit, where conventional analogue flight instruments are replaced by cathode ray tube (CRT) displays presenting a range of engine, flight and navigational data, together with a computerized FMS. This type of display is discussed in more detail in conjunction with another aircraft accident described in Section 12.3 in Chapter 12.

At the centre of the B737-400 panel is a CRT display presenting primary and secondary data relating to operation of the twin CSM56–3C turbofan engines. By convention the engines are numbered from the left facing forward, No. 1 being the left-hand engine and No. 2 the right-hand engine, being referred to by the flight crew as 'No. 1' and 'No. 2' respectively. The First Officer, who sits in the right-hand seat, was piloting the aircraft on this flight, rather than the Captain. The take-off and initial stages of the flight took place without incident until the aircraft was about 25 nautical miles (nm) south of Derby at an altitude of 28,300 feet. At about 20.05 hours GMT the No. 1, left hand, engine vibration indicator began to fluctuate and the pilots noticed a strong shaking of the aircraft frame. There were no other engine alarms or fire warnings showing at this stage. At the same time that this was happening, pale wisps of smoke began to enter the passenger cabin via the air-conditioning ducts accompanied by a smell of burning. Passengers on the port (left) side of the cabin noticed bright flashes coming from the No. 1 engine. When the engine malfunction was detected the Captain took over control of the aircraft from the First Officer, his first task being to disengage the autopilot. The Captain then spoke to the First Officer and asked which engine was giving trouble. After some hesitation, the First Officer replied that it was the No. 2 engine. The Captain then ordered him to throttle back the No. 2 engine. Following the Captain's instructions, the First Officer disengaged the auto-throttle for both engines (which controls the aircraft speed at a set value) and proceeded manually to throttle back the No. 2 engine to the idle setting. At this moment, the vibration of the airframe ceased but the vibration indicator on the No. 1 engine still indicated a vibration problem.

The aircraft was now flying on the No. 1 engine only. The First Officer called London ATC to declare an emergency due to a suspected engine fire and requested a diversion to the nearest airport. This was identified as East Midlands Airport near Nottingham. The aircraft was then cleared by ATC to descend to 10,000 feet. As a degree of panic had set in among the passengers, the Captain made an announcement over the public address system informing them that there was trouble with the right engine, which had now been shut down, and the aircraft was being diverted to East Midlands airport. After descent to a lower altitude, the aircraft was cleared to land at East Midlands airport, Runway 27. As the aircraft lined up for the approach the power to the No. 1 engine was further reduced and gradually the aircraft descended following the glide path. Just as the runway lights were coming into view about 2½ miles from touchdown, and at about 900 feet above the ground, the No. 1 engine suddenly lost power.

The Captain passed instructions to the First Officer to relight the No. 2 engine and at the same time attempted to lift the nose in order to extend the aircraft's glide onto the runway touchdown area which was by now fully in view. There was clearly insufficient time to relight the No. 2 engine and at this point the ground proximity warning alarm started to sound and the 'stick shaker' began to operate. This was the pilot's warning of an imminent stall condition. The aircraft rapidly lost height and hit the ground short of the runway at the edge of a field on the eastern side of the M1 motorway. As it hit the ground the main undercarriage was sheared off. Still travelling at 100 knots the aircraft demolished a fence, sliced through trees and having crossed all four carriageways of the motorway, collided with the far embankment in a nose-up attitude. The fuselage was fractured in two places while the tail section jack-knifed and came to rest in an upside-down position. The left wing struck a lamp standard fracturing a fuel tank and spilling aviation fuel down the embankment.

Fortunately there were no vehicles on the motorway at the exact moment of impact, but a driver who arrived at the scene a few seconds later was able to assist a number of injured passengers to evacuate the fuselage. The emergency services arrived at the scene in about 20 minutes and covered the area with a foam blanket to prevent ignition of aviation fuel leaking from the fractured wing tank. The accident was only survivable because the fuel which leaked from the aircraft did not ignite and cause a major fire. In the absence of fire it was possible to release injured passengers who would otherwise have been killed by fire and smoke.

Nevertheless, impact trauma caused the deaths of thirty-nine passengers while 76 passengers and seven crew members including both pilots, were seriously injured. Most of the fatalities occurred in the forward section of the fuselage which was badly crushed. Many of the injuries sustained by passengers were to the lower leg and foot as a result of failures of seat attachments to the floor and failures of seat restraints. These failures allowed seats to be thrown forward resulting in crushing or pelvic injuries. In the event of fire, the disabling effects of these injuries would have resulted in the accident becoming non-survivable for most passengers because of their inability to self-evacuate from the aircraft.

11.3.3 The causes

11.3.3.1 *The direct cause*

The direct cause of the accident was a failure on the part of the Captain and First Officer correctly to diagnose which engine was causing the problem. Engine failures are not uncommon and flight crew are not only trained to deal with such emergencies but are tested at regular intervals on the recovery procedures needed to prevent such an engine failure turning into an accident. They must pass these tests in order to maintain currency of their commercial licences. On this occasion, the failure was exacerbated by the initial failure of the First Officer correctly to identify the engine which was vibrating. It was this error that led to the shutdown of the engine that was performing normally.

The cause of the failure of the No. 1 engine was later found to be the fatigue failure of a turbine fan blade. The CFM56 engine type had been developed jointly by General Electric in the USA and SNECMA in France. The version in use on the B737-400 was a scaled-down, thrust-uprated engine which had previously been supplied for upgraded USAF Boeing KC135 military tankers. This type of engine failure had already occurred on a number of aircraft brought into service and had been the subject of occurrence reports. Two similar incidents which occurred shortly after the Kegworth crash showed that when a piece of fan blade became detached, the resulting out of balance allowed the rotor to come in contact with the fan casing, causing smoke, sparks and a risk of fire. The problems with the blades occurred when the engine was delivering rated climb power above 10,000 feet.

11.3.3.2 *The root causes*

An investigation into the root causes of the accident was conducted by the Air Accident Investigation Branch (AAIB) of the Department of Transport (AAIB, 1990). It concentrated on why the pilots had incorrectly identified the engine which was vibrating and as a result had shut down the unaffected engine. There were two important aspects to this error. First of all, why the error occurred and second, why error recovery did not take place. For the first aspect, the investigation directed its attention to the design and layout of the engine instrumentation, and how this might have misled the pilots into making the faulty identification. For the second aspect, it was obvious that the cabin occupants on the port side of the aircraft were aware that the sparks were coming from the No. 1 engine on the left and not the No. 2 engine on the right as reported over the public address (PA) system by the Captain. Attention was therefore paid to the social and cultural aspects of interactions between flight crew, cabin crew and passengers. Each aspect is discussed below.

11.3.3.2.1 Why was the wrong engine shut down?

The engine instruments

In order to understand why the wrong engine was shut down it is necessary to examine in some detail the engine instruments on the B737-400 panel and the ergonomics of their layout. The panel is a 'glass cockpit' comprising CRT displays with solid-state rather than analogue instruments. The display for the engine instruments is logically located in the centre of the console above the two thrust levers for the 'No. 1' and 'No. 2' engines. The engine instruments are split into two categories of primary and secondary instruments, the categories reflecting their relative importance to the flight crew when they make routine visual scans of engine instrumentation. The primary instruments which are displayed are as follows:

- N1 and N2, indicating the percentage of thrust being delivered by the fan and turbine rotors, respectively and
- Exhaust gas temperature (EGT) from the turbines.

The secondary instruments mainly indicate lubricating and hydraulic oil parameters including oil pressure, temperature and quantity. The secondary instruments also include the engine vibration monitors.

There is a convention used in the design of instrument displays which will be familiar to anyone who has operated an instrument panel. Where there are a number of identical equipment items, instrument readings relating to the same item of equipment are displayed in a vertical column with the name or identifier for the equipment item printed at the top or bottom of the column. Instruments which are measuring the same parameter will be displayed in a horizontal row with the name of the parameter indicated on each instrument. Whether using electro-mechanical or solid-state instruments displayed on a CRT, this convention will usually apply. Many solid-state CRT displays attempt to mimic conventional analogue displays in order to retain some familiarity for those used to reading electro-mechanical instruments. Thus, for the engine instruments on the B737-400 panel, there are two sets of parameters which need to be combined in the display, these being:

1. the primary and secondary engine instruments and
2. the corresponding displays of these instruments for Nos 1 and 2 engines.

The problem then becomes how best to integrate these within the display to present the information in the most intuitive and user-friendly way.

Alternative layouts of displays
Alternative layouts of the engine instruments are shown in Figure 11.3 and are indicative of the way the readings on a B737-400 aircraft could be displayed. The alternatives would be applicable not only to solid-state displays, as on this aircraft, but also to conventional analogue instruments; the ergonomic principles relating to readability would be exactly the same.

Layout option A – as on Boeing 737-400 aircraft
In layout option A, all the primary instruments for both engines are grouped together in a box on the left, with the 'No. 1' and 'No. 2' engine instruments situated to the left and right, respectively within this display box. Secondary instruments are similarly grouped in a box but located to the right of the primary instrument display. The difference in size between primary and secondary instruments shown reflects the difference in their importance. This was the layout adopted on the B737-400 involved in the accident at Kegworth. For each set of instruments, the user must always remember that within the

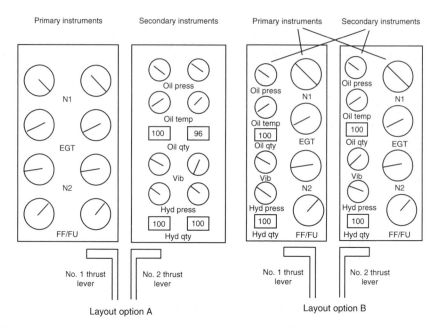

Figure 11.3 Alternative layouts of engine instruments for B737-400 panel.

left-hand box, the No. 1 engine instruments are on the left, and the No. 2 instruments are on the right. This is in spite of the fact that *all* the primary instruments, located on the left, are placed above the No. 1 engine thrust lever, and *all* the secondary instruments are located on the right above the No. 2 engine thrust lever. The danger thus arises that a No. 2 engine *primary* instrument could be associated with the No. 1 engine and a No. 1 *secondary* instrument might easily be associated with the No. 2 engine due to the positioning above the thrust levers. To associate a primary instrument with the No. 2 engine it is necessary to think 'left and right' and ignore the position of the No. 2 engine primary instruments above the No. 1 thrust lever. In the same way, to associate a secondary instrument with the No. 1 engine it is also necessary to think 'left and right' and ignore the position of the No. 1 secondary instruments above the No. 2 thrust lever.

Whether a primary instrument can be mentally associated with the correct engine is dependent upon whether the effort in differentiating between left and right (called 'handedness', see below) is overridden by the strength of the intuitive association between primary instruments and the No. 1 engine thrust lever below. The same situation exists for the secondary instruments, except in this case the competition is between left and right differentiation and the strong association of these instruments with the No. 2 engine thrust lever. It should be noted that the vibration instruments for the Nos 1 and 2 engines are in the secondary grouping and are therefore above the No. 2 thrust lever. Referring to the B737-400 layout A in Figure 11.3, it can be seen how the high vibration read-out on the No. 1 engine, could easily be mistaken as a high vibration on the No. 2 engine if the association with the No. 2 thrust lever overcame the left/right differentiation of the user.

Layout option B

Layout option B in Figure 11.3 groups both primary and secondary instruments together above the engine thrust lever with which they are associated. This layout was not adopted in the B737-400 aircraft, but was suggested in the accident report as a more user-friendly alternative. With this layout there is much less chance that a No. 1 engine instrument could be misread as a No. 2 engine instrument or *vice versa*. It should also be noted that the relationship between primary and secondary instruments for Nos 1 and 2 engines is identical and not a mirror image. This also aids consistent and correct association of instruments with the engine to which they relate. The ergonomically inferior layout option A that was designed into the instrument panel of the B737-400 aircraft certainly contributed to the misdiagnosis of the problem by the pilots. The error almost certainly had a systemic cause associated with a confusing layout of the engine instruments rather than, for instance, carelessness. The systemic cause is also acknowledged by the accident report.

Other factors contributing to the error

It needs to be remembered that the Second Officer was required to identify the vibrating engine at a time when both workload and stress levels in the cockpit were high. Attentional resources are always more limited when workload is high and at such a time a principle of cognitive economy is called upon as described in Chapter 2, Section 2.3.2.1. What are perceived to be the more important tasks are given a higher priority at the expense of less important tasks. The error probability for any of these tasks will increase compared with what might be expected if a task was carried out on its own. Along with the need to identify which engine was vibrating, it was also necessary for the pilots to pay attention to the cause of the vibration and assesses whether its severity would demand a shutdown of the engine. If a shutdown is delayed then the probability of an engine fire is increased, but if it is left running and throttled down, it still has limited availability. On top of all this it is necessary to take over manual control of the aircraft from the automatic pilot, assess the location of the aircraft and decide whether to return to the airport of departure or to an alternative. All these matters are competing for the limited resources of the pilots.

Once the misdiagnosis error had been made, it became more likely that the decision on which engine was causing the problem would not be questioned again unless evidence that contradicted it was forcibly brought to the attention of the flight crew. In this respect, the active error can be classed as leading to the type of scenario known as a 'mistake', or planning error. A 'mistake' occurs when an inappropriate plan is formulated even though it may be executed correctly. It usually happens when a plan is predicated on a wrong assumption or misleading information. During this incident, there were many opportunities for recovery from the initial error and these are discussed below.

11.3.3.2.2 Why did error recovery not take place?
The mindset in the cockpit

Once the failed engine had been identified incorrectly as the No. 2 engine, it appears that the scenario in the cockpit was inevitably set for disaster. It led to a syndrome referred to as a 'mindset', one that has long been recognized as the cause of many accidents. Its effects are most often seen in the area of diagnosis and decision-making. Once a diagnosis or decision has been made, particularly at a time of great stress or when excessive demands are placed upon cognitive resources, then the more time that elapses, the less likely it is that the reasoning will be re-evaluated. This is especially the case where no later information is received in the form of visual, audible or other cues which would lead to the diagnosis being questioned. Indeed, it is quite possible that there will be bias so that information received later is interpreted in a way that supports the original diagnosis. Again, this follows the principle of 'cognitive economy'; to question the original thinking at a later stage may be seen as costing valuable time and resources. The imperative is to solve the problem, not to re-think it without a compelling reason to do so. In this accident there was information available to the flight crew which was used as 'evidence' in support of the mindset, but there was also evidence which worked against it and unfortunately this was not made available. The information is summarized below.

'Evidence' supporting the mindset

There were a number of factors reinforcing the mindset that the No. 1 engine had failed, causing it not to be questioned. These were:

1. When the No. 2 engine was shut down, there appeared to be a reduction in the level of vibration of the aircraft in spite of the fact that the No. 1 vibration gauge continued to show the same maximum reading. This reinforced the mindset by confirming to the flight crew, at an early stage that the diagnosis of a No. 2 engine failure was correct. During the investigation it was revealed that this reduction in vibration could have been expected as a result of disconnecting No. 1 engine from the auto-throttle. The reason was that the engine, on being disconnected from the auto-throttle, immediately recovered from a series of compressor stalls which were adding to the vibration.

2. In the course of investigation it was discovered that the Captain had a view that engine vibration indicators were generally unreliable and should not be depended upon. This was based upon previous experience in flying older DC9 twin jets. He was not aware that technical developments had since led to a much more reliable design of instrument. There was no evidence that the vibration instruments on the B737-400 aircraft were unreliable since they operated on an entirely different principle to the older types.

3. During the investigation the Captain said he gained support for the decision that the problem was on the No. 2 engine because this engine supplied air-conditioning to the passenger cabin from where smoke had been reported. In fact, the air-conditioning was also supplied from the No. 1 engine but the Captain was unaware of this. In addition, the Captain was unaware of smoke in the passenger cabin when the decision was taken to shut down the No. 2 engine. It is possible that his recollection was coloured by information received after the accident.

The above factors were 'supporting evidence' propping up the mindset. As more 'evidence' was accumulated, the strength of the mindset was increased and the chance of recovery was reduced.

'Evidence' against the mindset which was not made available

Other information, apart from that available to the flight crew in the cockpit, and which could have caused a re-examination of their decision to shut down the No. 2 engine was available but was not transmitted to them. A number of passengers in the cabin had observed sparks coming from the left-hand engine and were convinced that this was where the problem lay. Some of them queried the cabin crew on why the Captain in his broadcast had referred to shutting down the right-hand engine. However, this information was not relayed by the cabin crew to the Captain or First Officer. Whether it would have caused them to review their decision cannot of course be known with certainty. The fact that the information was not transmitted was at least in part caused by a hierarchical culture that prevails within the aviation industry. The Captain of an aircraft is an authority figure who is usually regarded with some awe, certainly by the First Officer and cabin crew and possibly by the passengers. To question the decisions of the Captain, and interrupt him with what may be trivial queries when he is handling an emergency, is not something to be approached lightly. In the case of the cabin crew, this could risk a possible reprimand and in the case of passengers, ridicule or anger.

It may seem hard to believe that matters of hierarchy could interfere with the transmission of important and vital information in an emergency. Yet there are strong social and cultural factors adversely influencing personal interactions between authority figures, who are perceived as very powerful, and subordinates or members of the public. As far as the public is concerned, social barriers such as these have tended to break down in recent years, and there is now much more of a perception that authority is something that must be earned and justified rather than accepted without question. Nevertheless, a hierarchical culture that can hinder the free flow of information is known to exist in the aviation industry. It probably stems from a long history of recruiting commercial pilots from the ranks of retired military pilots. The rules of interaction between superior and subordinate, although not quite as rigid as in the armed services are still governed to some degree by the same rules. Airlines are a uniformed organization and this tends to reinforce the authoritarian obedience to instructions.

This culture has long been recognized as the cause of a number of serious accidents to commercial aircraft including the world's worst aircraft accident in terms of fatalities which occurred in March 1977 at Las Palmas airport, Tenerife, in the Canary Islands. The accident was principally caused by the failure of a First Officer to question seriously the actions of the company's most senior training pilot who was acting as Captain of a KLM B747. The aircraft was awaiting take-off clearance at the runway threshold. Due to fog he was unable to see that a Pan Am B747 was backtracking towards him along the same runway. For reasons that are unclear, the Captain of the KLM aircraft made a decision to initiate the take-off without proper clearance from ATC. The result was a collision with the Pan Am aircraft that led to 583 fatalities. Conversations recorded from the cockpit show that although the Second Officer queried the actions of the Captain, he was immediately rebuffed and made no further comments in spite of his serious reservations.

After the accident at Las Palmas, the hazard of poor communication between the different hierarchical levels on commercial aircraft was considered to be so serious within the industry that it led to the widespread adoption of a technique known as crew resource management (CRM). CRM aims not only to enhance the awareness of such problems among aircrew, but also trains every level of staff to take full advantage of all resources available to them, including information from cabin crew and passengers. A similar problem appears to have occurred on the aircraft involved in the Kegworth disaster. The AAIB report into the accident comments that 'the pilots of this aircraft did not make effective use of the cabin crew as an additional source of information' (AAIB, 1990). The report also notes that pilot training in the UK concentrates mainly on two aspects, these being:

(a) aircraft handling skills involving psychomotor aspects,
(b) dealing with specific emergencies using a procedural or rule based approach.

Although both aspects are essential elements of flight training, it is rare that an actual emergency can be dealt with by following a specific procedure to the letter. All emergencies will have unique features which are not easy to plan for in advance. As discussed in Chapter 2, diagnosis and decision-making are knowledge based, rather than rule based activities, and as a result are subject to forms of human error which may pass unrecognized. The AAIB report notes that once the decision was made that the problem was in the No. 2 engine, the pilots carried out an almost faultless engine shutdown procedure. However, the report recognizes that the failure lay not at the rule based level but at the strategic or decision-making level involving mental formulation of an appropriate plan to deal with the emergency. Training in knowledge or decision based activities is not provided to pilots. Pilots may therefore be unaware of the pitfalls awaiting them when a plan is prepared 'on the hoof' at a time of high workload and stress. This subject is addressed in more detail in Chapter 13, Incident Response Errors, using as a case study the Swissair flight SR777 accident in Nova Scotia in 1998.

11.3.4 Conclusion

11.3.4.1 *Laterality or handedness*

Directionality is defined as a person's ability mentally to project direction in space including, for instance, the ability to know north, south, east and west from the present position or indeed the ability to look at another person standing face-to-face and be able to identify which is their right hand. Laterality is more specific and is defined as a person's internal awareness of up and down and left and right. Handedness more specifically refers to the ability to distinguish between left and right and in particular to be able to co-ordinate eye and hand in response to this knowledge. This ability clearly varies between individuals and is generally considered by psychologists to be innate. Although everyone will from time to time fail to distinguish left from right, some people are found to be more prone to this than others. It is has been found for instance that left-handed people or people who are even slightly dyslexic may be more prone to the problem. It is also found that laterality problems may increase under conditions of stress or high workload. Indeed, as was the case with the Kegworth accident, the problem of laterality and handedness can be worsened as a result of poor ergonomics or confusing visual cues.

It might be expected that the selection of pilots would take account of this. However, it is not the case and as far as the author is aware, no psychological testing is currently employed by airlines to identify a tendency to poor laterality. A psychological study of pilots for the Norwegian Air Force (Gerhardt, 1959) found that that laterality problems were not uncommon among the group of pilots studied. A number of pilots had difficulties in identifying left and right, including one pilot who could only identify his left hand by looking for the hand which was wearing a wedding ring. Some of the pilots had a tendency to place sequential numbers in the wrong order and demonstrated uncertainty regarding the mental projection of direction in space. No evidence arose from the investigation of the Kegworth accident that the two pilots had any sort of laterality problem, but the accident may indicate that prospective airline pilots should be psychologically screened for these tendencies. The fact that laterality problems are made more probable under stress or high workload must also be take into account in the design of cockpit layouts. This is a topic in its own right but is briefly discussed below in the context of the Kegworth accident.

11.3.4.2 *Cockpit ergonomics*

The scanning of engine instrument readings (together with other flight instruments) for abnormal or out-of-range values, such as rotor speed, exhaust gas temperature, vibration, low oil pressure or high oil temperature, is a routine activity conducted by pilots during flight. There was a time when a third person was employed in the cockpit as a flight engineer with specific responsibilities for monitoring engine performance. However, with the widespread adoption of jet and turbine technology major improvements were obtained in the reliability of aircraft engines and the services of the flight engineer were dispensed with on economic grounds. In addition it was judged that the greater use of FMS for most of the flight took sufficient workload away from the pilot and co-pilot to allow a two-man crew to monitor engine performance as well as fly and navigate the aircraft. The reduction of staffing in the cockpit does not, of course, necessarily take account of the increased workload when an emergency occurs. This problem is further discussed in Chapter 13.

When the reading from a primary instrument is abnormal or out-of-range it will normally be brought to the pilot's attention by a visual/audible alarm. Thus the pilot is not entirely reliant on a visual scan. In the case of the aircraft involved in the Kegworth accident, it was a secondary engine parameter (the vibration monitor), which was out of range. Since this was deemed to be a less important instrument, the solid-state display which showed the reading on the CRT screen was much smaller than the primary displays. Furthermore, the moving indicator which registered the reading was a very small light emitting diode (LED) pointer moving round the outside of a circular scale. By comparison, the mechanical pointers on the older type of analogue instrument were much larger and clearer and their position on the display was unambiguous. With the CRT display there was no audible or visual alarm to indicate a divergence from normal. This is possibly because the provision of too many alarms can increase confusion and it was necessary to be selective. However, if such an alarm had been provided it might have been colour-coded in some way to distinguish between left and right engines.

Investigation into the accident revealed that pilots from this particular airline had only been given one day's training to convert from the B737, Series 300 (with the older type of cockpit instruments) to the B737, Series 400 aircraft (with a 'glass cockpit') prior to

Table 11.2 Qualitative human error analysis of the Kegworth accident

Cause	Description	Type of error or other failure	Systemic cause or preventive action
Direct cause	Failure correctly to identify the engine which had failed	Active error followed by a 'mistake'	Flight crew training in high-level decision-making (how a 'mistake' can follow a faulty diagnosis).
Root cause	Poor ergonomic layout of cockpit display instruments (initial error)	Design error	Designers to be aware of potential problems of laterality and handedness especially at times of high workload.
	Failure in communication with cabin crew (recovery error)	Management error	Cultural hangover from early days of flight where the Captain is an authority figure not to be questioned. Cockpit Resource Management training aids recognition and provides solutions to this problem.
Contributory causes	Minimal training on 'glass cockpit' of B737-400 series	Management error	Familiarization training on a simulator especially in handling emergencies.
	Poor readability of glass cockpit instruments	Design error	Provision of audio/visual alarms for out-of-range values where readability is poor.
	Possible problems of laterality of co-pilot	Active error	Psychological screening of aircrew candidates.

taking command. The ability to deal with the emergency may have been hindered by unfamiliarity with the cockpit layout and displays. It is certainly unacceptable that familiarity should be gained during a real emergency. The problem was exacerbated by the fact that there was no flight simulator available for pilot training that was equipped with the new instrumentation. Had the flight crew been presented with a more familiar display it is possible that this accident might not have happened.

11.3.4.3 *Human error analysis*
The qualitative analysis in Table 11.2 summarizes the main direct, root and contributory causes of the accident as discussed above.

References

Air Accident Investigation Branch (1990). *Report on the Accident to Boeing 737-400, G-OBME near Kegworth, Leicestershire, on 8th January 1989*, London: Department of Transport (Aircraft Accident Report No. 4/90 (EW/C1095)) available on the Internet at http://www.aaib.dtlr.gov.uk/ formal/gobme/gobmerep.htm

Gerhardt, R. (1959). *The Human Factor in Aviation*, Oslo: Norwegian Armed Forces Publication.

International Civil Aviation Organization (1993). *Destruction of Korean Airlines Boeing 747 on 31st of August 1983. Report of the Completion of the ICAO Fact-finding Investigation*, Montreal: ICAO.

12

Violations

12.1 Introduction

The distinction between violations and human errors is described in some detail in Chapter 3, Section 3.3.3. The most important difference was found to be one of intention. By definition, it was decided for the purposes of this book that human errors are by nature unintentional or inadvertent. On the other hand, a violation is, by definition, always carried out with conscious intent with or without knowledge of the consequence. As with an error, recovery from a violation is possible by a subsequent action that prevents or mitigates the consequence. It is described in Chapter 3 how violations may be classified on the basis of frequency, such that there may be exceptional violations (occurring only rarely) or routine violations (occurring on a regular basis and possibly becoming custom and practice). Violations can also be classified on the basis of causation, and in a similar way to errors, a classification of skill based, rule based or knowledge based violations can be applied, depending upon the circumstances in which the violation occurred and the type of 'rule' that was violated.

Two case studies from the nuclear and the aviation industries are presented in this chapter to illustrate violations that have led to major accidents. The Chernobyl accident that occurred in the Soviet Union in 1986 was the world's worst nuclear accident in terms of its immediate and long-term consequences. The accident received headline coverage worldwide and made the Western nuclear power industry closely scrutinize the safety of their operations. The legacy of the accident in terms of radiation related sickness is still present. It was caused by a violation at the rule based level during the planned test of a reactor shutdown system. It is interesting because the violations were of existing rules yet formed part of a test procedure managed by senior engineers at the plant. The case study examines how a violation of safety rules could take place in a highly bureaucratic rule based environment. There are similarities to the Kegworth accident in the previous chapter since there was a group mindset which, in this case, judged that a novel situation justified violation of the normal rules.

The second case study examines a much lesser-known aircraft accident which took place at Mulhouse in France in 1988 involving the crash of an Airbus A320 aircraft with 130 passengers on board during a demonstration flypast at a small aerodrome. It was one of a number of accidents which occurred around the time when new and sophisticated computerized flight management systems (FMS) had been introduced into a range of Airbus aircraft. The violations which led to this accident involved two senior training pilots concurring to carry out a dangerous low-level manoeuvre in an aircraft with which they were not completely familiar. Due to their over-confidence in the new technology, they were motivated to breach a number of fundamental flying rules which are part of the basic training of any aviator. Their actions were not therefore those that would be expected of a reasonably competent and experienced person. This brings the violation securely into the knowledge based classification described in Chapter 3.

12.2 The Chernobyl accident

12.2.1 Introduction

The world's worst nuclear accident at Chernobyl in the Soviet Union in 1986 had a major impact on the civil nuclear power industry worldwide. Its effects are still being seen in terms of radioactive contamination of land and incidence of cancers among downwind populations. The type of reactor involved in the accident was unique in the sense that when operating at low power it became unstable and was liable to go out of control. Very strict procedures were in place to prevent operation in the unstable regime, yet at the time of the accident many of these rules were violated in the interests of carrying out an experiment. This case study explores the weaknesses in the reactor design that allowed the accident to be triggered by a number of violations of operating rules. It describes how the violations were implicitly condoned by management and motivated by a group mindset that had developed in the control room.

12.2.2 The Chernobyl reactor

Construction at the site of the Chernobyl nuclear power station in the Ukraine (then in the Soviet Union) began in 1975. By 1986, the facility comprised four 1000 megawatts RBMK type nuclear reactors in operation with two more under construction. The plant was located about 60 miles north of the city of Kiev close to the town of Pripyat which at that time had a population of 49,000. The reactors were constructed in pairs and shared common utilities and buildings. The RBMK type of reactor is a graphite moderated, light water cooled design, superficially similar to the Magnox reactors operating in the UK, but much larger in size and power output. There are also major differences in the operating characteristics of the reactors and in the way they are operated and protected. A schematic of the design of the RBMK type of reactor is shown in Figure 12.1.

The RBMK reactor comprises a calandria (or steam-generating vessel) in which are placed zirconium alloy pressure tubes through which water is pumped. Within each tube

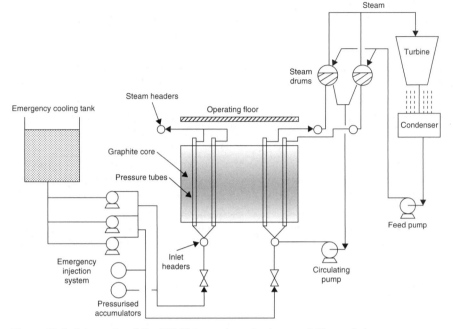

Figure 12.1 Schematic of the RBMK type of reactor in use at Chernobyl.

there is a uranium fuel element. Heat is removed from the fuel element converting water into steam collected in the steam drums. The superheated steam produced is passed to an electricity-generating turbine. Steam passing from the turbine is condensed and fed back into the steam drums using a feed pump.

The pressure tubes are surrounded by graphite blocks perforated by vertical channels, each channel containing a tube. Each pressure tube comprises 36 separate fuel rods in an assembly 10 metres long. The separate fuel rods are made up of 2 per cent enriched uranium oxide pellets contained in zircaloy tubing.

The graphite core weighs 1700 tonnes and is surrounded by helium and nitrogen. The purpose of the graphite is to moderate or slow down the neutrons produced by fission of the uranium thus enabling an efficient chain reaction to be maintained. In order to prevent a runaway chain reaction and to control the reactivity of the core, the reactor is controlled by 211 boron carbide control rods which can be raised or lowered in special channels in the graphite and cooled by water from a separate circuit. The purpose of the rods is to absorb neutrons thus controlling the amount of fission taking place. This enables:

- automatic control of the power output from the reactor,
- rapid power reduction,
- 'scramming' of the reactor in an emergency using additional protection rods.

An automatic reactor trip drops these rods into the reactor core and can be caused by many events such as low water level in the steam drum. Above the reactor there is a biological radiation shield comprising a steel and concrete pile cap 3 metres in thickness penetrated by holes through which fuel rods can be removed and replaced by a refuelling machine. The RBMK reactor has no secondary containment to retain radioactivity in the event of a breach of reactor integrity. When the reactor is shut down it is necessary to remove residual heat to prevent a temperature rise in the reactor which could damage the core. This is provided by an emergency core cooling system (ECCS) that ensures a flow of cooling water into each fuel element using injection pumps and pressurized accumulators charged by a separate pumped circuit. The ECCS also provides cooling of the core in the event of a pipe failure in the main boiler circulating system.

Following a reactor trip, the emergency protection systems require a source of electrical power with a break of only a few seconds allowed. When the reactor is tripped, the residual kinetic energy of the still rotating turbo-generator will supply this power for a short period. In April 1986, a test was planned to establish the time that power could still be supplied to the emergency systems after a reactor trip. An earlier test had been unsuccessful due to inadequate control of the generator magnetic field. A new field controller had been fitted and the new test was planned for 26 April 1986.

12.2.3 Problems with the design

The Soviet authorities and other nuclear operators around the world were aware of a major problem with the RBMK reactor design. The problem concerns a characteristic known by nuclear engineers as a 'positive void coefficient'. This means that if there is an increase in power from the fuel or a reduction in the flow of water past the fuel rods (or both), then the presence of steam bubbles in the fuel channels increases. Since water absorbs more neutrons than steam this leads to an increase in the number of neutrons present and an increased number of fissions in the uranium fuel. This causes an increase of reactor power output. An increase in power also causes a rise in fuel temperature that has the opposite effect and decreases the neutron population. The balance between these two effects depends almost entirely upon the power output level from the reactor at the time.

At normal levels of power generation, the fuel temperature effect overrides the effect of steam bubbles in the tubes, and the 'void coefficient' is said to be negative. However, if the reactor output falls below about 20 per cent of full power, then the coefficient can become positive and reactivity of the core will increase with the danger of a runaway nuclear reaction. As reactivity increases, more steam bubbles are produced, increasing reactivity, and so on until the reactor becomes unstable. The effect takes place extremely rapidly and in less than the 20 seconds required for operation of the shutdown system. This 20 second response of the reactor control and shutdown system to a demand is very slow compared with other designs of reactors and is an additional problem for the RBMK type. For this reason the management of the Chernobyl reactor had prohibited operation below 20 per cent power.

Another characteristics of the RBMK design is that when the absorber rods are inserted into the reactor to reduce power, the initial effect is to displace water from the core and cause a sudden increase in power. This effect is known as 'positive scram' and was another factor in the Chernobyl accident. In the 1970s and 1980s the RBMK type had been considered by a number of countries who were developing their own nuclear power programmes, but because of these shortcomings the design was never adopted outside the Soviet bloc.

12.2.4 The accident

12.2.4.1 *The turbo-generator test*

The accident at Chernobyl took place as a result of a test carried out on 26 April 1986 in Unit 4 of the power station. The test was designed to check whether, in the event of a total loss of electrical power, the rundown of the turbo-generator would produce sufficient electricity to keep the safety systems operable for a period of 40 seconds. This was the time taken by the backup diesel generators to supply emergency power.

A programme for the test had been supplied by the manufacturer of the electrical generators and their personnel would be present. The test was to be overseen by the Deputy Chief Engineer of Units 3 and 4. The operators were required to bring the reactor to low power before shutting it down and allowing the turbine generator to run down and supply electricity to the safety systems. At the time of the test the plant had been scheduled to undergo routine maintenance and it was decided to carry out the test as part of the shutdown procedure.

The test sequence commenced at 01.00 hours on 24 April and by noon on the following day the reactor output was down to 50 per cent. At this time the ECCS was disconnected in direct violation of standing procedures. However, because of a grid requirement for power, it was requested that further reductions in power were postponed until 23.00 hours on the evening of the 25 April. The reactor continued to be operated in violation of procedures with the ECCS disconnected.

At 00.00 hours on the 26 April, the nightshift came on duty and preparations were made to continue the test. At this point the test procedure required the control rods to be changed from automatic to manual operation, but as a result of doing this, the reactor output suddenly experienced a transient and fell dramatically to about 30 megawatts due to instability. Thirty megawatts output was a dangerously low level of operation in view of the negative void coefficient of the reactor. The power was gradually brought up to 200 megawatts , about 20 per cent of normal, by 01.00 hours. This was thought to be a safe level, but unknown to the operators, poisoning of the reactor fuel rods with xenon (a fission by-product) had occurred as a result of the transient. This had the effect of reducing the safe margin of reactivity of the reactor to below that required by the regulations governing operation.

At 01.07 hours, two reserve water circulating pumps were switched on, another prohibited practice due to the danger of pump cavitation, a phenomena whereby bubbles of steam are entrained into the water due to a drop in suction pressure. The result of this was a decrease in the amount of steam being generated and a drop in pressure which affected reactivity due to the positive void coefficient. The water level in the steam drums now dropped to below the reactor trip setting. This would normally have scrammed the reactor but the operators had already overridden the protective systems in preparation for the test in further violation of the operating procedures. In spite of the unstable operating regime now prevailing, the planned test was continued and one of the two turbogenerators was disconnected and shut down. The electrical supplies for the emergency systems were connected to the other generator.

At 01.23 hours the steam stop valve to the No. 8 turbo-generator was closed and the generator allowed to run down. Due to the reduction in steam flow from the drums, the pressure in the steam system started to rise and as a result the flow of water through the pressure tubes began to fall as the back-pressure increased on the circulating pumps. This coincided with a reduction in power being delivered to the circulating pumps due to the unloading of the No. 8 turbine generator. At this point the reactor went 'prompt-critical', a condition where the reactor goes critical on prompt neutrons alone, and no longer requires the effects of a moderator to slow down prompt neutrons to thermal neutrons. Under prompt critical conditions, the time between nuclear fission generations drops to one-millionth of the time between fissions in normal operation. As a result the reactor cooling system cannot absorb the heat from the core and this leads to immediate 'flashing' of water to steam exceeding the design pressure of the steam system. A prompt-critical reactor is impossible to control, since the response of the control system would have to be measured in nanoseconds. The operator, realizing that the power from the reactor was rising exponentially decided to activate the emergency shutdown. This was far too late to have any effect since the reactor at Chernobyl was already out of control.

12.2.4.2 *The aftermath*

The surge in power from the reactor caused the fuel to increase in temperature until it reached its melting point when it started to disintegrate. Fragments of molten fuel were discharged into the surrounding water jacket causing a series of steam explosions rupturing the tubes and blowing the pile cap off the reactor. Almost immediately the containment was breached and radioactive materials and burning graphite blocks were thrown into the air. At the centre of the reactor, the remains of the burning graphite moderator together with radioactive fuel particles formed a white-hot glowing mass emitting gamma and neutron radiation. Water contaminated with radioactivity flowed from the reactor onto the surrounding area. Over the days that followed about 5000 tonnes of inert materials were dropped from helicopters into the centre of the reactor to mitigate the release of fission products. The materials included limestone to generate carbon dioxide to extinguish the flames and boron carbide to shut down any remaining fission processes. Due to the insulating effect of these materials, the reactor core temperature rose over the next few days to achieve nearly 2000°C and further releases of radioactivity occurred, particularly iodine.

Radioactive releases were initially transported in a north-westerly direction towards Scandinavia, but later as the winds turned easterly, fission products were swept across Western Europe reaching the UK around 2 May. In spite of the severity of the accident, it is estimated that the average individual dose received in the UK in May, integrated over a 50-year period represented only about 1–2 weeks of normal background radiation or the equivalent of a '3 week holiday in Cornwall' (Gittus *et al.*, 1988). Between July and November 1986, a concrete and steel sarcophagus was built to entomb the remains of Reactor No. 4. This was hurriedly built under adverse conditions and has an estimated lifespan of only 30 years. Within a few years the structure had already started to deteriorate and needed to be reinforced. In February 1997 international funding was obtained to strengthen the structure and remove radioactive dust and other material for disposal. Although now safe, the sarcophagus will eventually have to be replaced.

Twenty-six deaths occurred on site at the time of the accident as a result of radiation poisoning and four site workers died as a result of burns or falling masonry, making a total of 30 fatalities. An increased incidence of thyroid cancers, usually treatable and generally non-fatal, occurred in the region of Belarus at levels of 100 cases per million in children under 15 years. The comparable incidence in the UK is about 0.5 per million on average. This is probably linked to emissions of radioactive iodine following the accident.

The present-day cost to the Republic of the Ukraine in dealing with the consequences of the Chernobyl accident amounts to nearly 12 per cent of its national budget. (International Atomic Energy Agency, 1997). Following the accident, the town of Pripyat was evacuated and by 4 May a total of 116,000 people had been evacuated from a 30 kilometre exclusion zone. About 50 million Curies of radioactivity were released during the accident. The release was 1000 times more than the Windscale fire in the UK in 1957 and 1 million times more than the Three Mile Island accident in the USA in 1979.

12.2.5 The causes

12.2.5.1 *General*

The root cause of the accident was a design deficiency of the reactor which allowed a test procedure to take reactor conditions beyond the safe operating envelope. Although the test procedure was planned in advance, it involved the violation of a number of normal operating rules specifically designed to protect the reactor against such an event. These violations were therefore the direct cause of the accident. There were a number of opportunities during the procedure where the reactor could and should have been shut down safely had the existing safety rules been observed. However, for reasons which are discussed below, the completion of the test procedure took precedence over the normal rules which were comprehensively ignored.

The plan that had been formulated to carry out the test was itself fundamentally flawed. The same experiment had been carried out on two previous occasions in 1982 and 1984 on the same type of reactor, although not at Chernobyl. On both occasions, a single turbo-generator was allowed to run down in an attempt to keep the circulating pumps operating for a few tens of seconds, the time needed for an emergency electrical supply to come on-line. The tests were not successful but a dangerous situation was avoided on these occasions because the reactor was tripped in advance of the test being carried out and was therefore protected against instability. The flaw in the formulation of the plan for No. 4 Reactor at Chernobyl was that the reactor was not tripped prior to the test and the situation was made worse because the reactor was already in an unstable operating mode when the experiment was commenced.

12.2.5.2 *The design faults*

The design deficiencies of the reactor are discussed in some detail in Section 12.2.3. The management of this and other power stations operating the RBMK reactor type were fully aware of the positive void coefficient and other problems that could lead to reactor instability at low power output. As a result the operating procedures were designed to ensure that the reactor was never operated in an unstable regime. Chapter 13 includes two case studies involving errors in incident response. In each case, attempts were to compensate for an inherently unsafe design by the use of operating procedures. The use of procedures to compensate for an uncorrected design fault is nearly always unsatisfactory. It effectively means that the prevention of an accident will depend, unrealistically, upon the absence of human error or rule violations.

12.2.5.3 *The violations*

This section identifies the violations (and some errors) that occurred at Chernobyl and explores the underlying reasons. Violations and errors that took place are described in terms of the events leading up to the disaster which are presented here in chronological order.

12.2.5.3.1 The emergency core cooling system

The ECCS was disconnected on the day before the test was carried out in accordance with the experimental test plan. The reason was to ensure that emergency cooling was not activated since it would interfere with the results of the test. It was required that the circulating pumps alone would supply water to the pressure tubes while the generator was running down. In the event, the ECCS would not have had an influence on the outcome of the accident had it been available. However it was a violation of the rules that the reactor was operated for nearly 12 hours with the ECCS disconnected. On the day previous to the test it was clear to the operators that a number of other normal safety rules would have to be disobeyed when the test was carried out. As a result, it is possible that a frame of mind developed among the operating team that in this situation the normal rules no longer applied. This perception was probably reinforced by the presence of a senior engineering manager who was to oversee the test, the presence of the manufacturer's representatives and a written procedure for the test that appeared to override the normal operating rules. In these circumstances, the violation by the plant operators of the ECCS rule and other rules later in the test becomes more understandable.

12.2.5.3.2 The reactor power reduction

During the procedure the operators were instructed to switch the control rods from automatic to manual. In making the changeover an error occurred and the reactor power output fell considerably below the minimum 20 per cent of full power stated by the rules. This was an operating error rather than a violation. However, according to the operating rules the reactor should have been tripped by dropping the control rods to bring it to a safe condition and the experiment brought to an end. Instead, a decision was made by the Deputy Engineering Manager to keep the reactor running, raise the power output again and continue operation so that the test could be completed. This further violation of the rules may have been motivated by a feeling that the test procedure had by now taken priority rendering the normal rules irrelevant for the period of the test. It is of interest that when the error was made the operators were criticized by the manager. When they were instructed to increase reactor power they raised objections but these were dismissed by the manager. Up to this point it is possible that the operators were still working in a rule based mode while the manager was operating in knowledge based mode with his mind fixed more on the completion of the test than on the observance of normal rules.

When the reactor was stabilized at 200 megawatts (about 20 per cent full power), xenon poisoning of the fuel had occurred and was only possible to maintain this power level by removing control rods from the reactor. Reactor operation in this mode was well outside the allowable limits of normal operation. The RBMK reactor was unique in another way. Reactors in the West have considerable diversity in shutdown systems. The RBMK reactor only possessed a single safety system capable of shutting the reactor down in an emergency. If a fault arose in this single system, a serious accident could ensue. The fact that xenon poisoning had occurred now made the reactor even more difficult to control. Nevertheless it was decided to proceed with the test.

12.2.5.3.3 The additional circulation pumps

An additional circulation pump was started in each of the main cooling circuits making four pumps in operation for each circuit. The intention was to power one of the circuits separately during the rundown of the turbo-generator. Apart from the fact that running four pumps would cause cavitation, it was not appreciated that the increased flow of water to the steam drums would depress steam generation and cause a fall in the level of the water in the steam drums.

This drop in level should have caused the reactor to trip but already the trip had been overridden in violation of the rules. To restore the level in the steam drums, the operators added water using the feedwater pumps. This reduced steam generation even further. The loss of voidage due to fewer steam bubbles in the pressure tubes, reduced the reactivity still further below the 20 per cent minimum, and the operators compensated for this by withdrawing more control rods manually. This meant that when the voidage increased again with steam production, there would be insufficient control rods inserted to control the increase in reactivity. Thus the scene was set for the unfolding disaster.

12.2.5.3.4 Rundown of the turbo-generator

Preparatory to the rundown of the turbo-generator, the steam bypass around the turbine was closed. This caused the steam pressure to increase and the level in the steam drum to rise again. As a result, the operators decided to reduce the quantity of feedwater being added. The voidage in the pressure tubes was now beginning to increase with steam production and since water absorbs more neutrons than steam, this led to an increase in the number of fissions and a rise in reactivity.

In order to monitor the situation it was decided to print out what is called the 'reactivity reserve'. This is a measure of the number of control rods that are held in reserve for insertion into the core to shut down the reactor. If control rods have already been withdrawn, then the reserve will be lower. It is also a measure of the operational safety margin of the reactor. It was found that this reserve was only six to eight rods and the normal rules required that the reactor should be shut down immediately. Unlike Western reactors, which would be shut down automatically on reaching such a dangerous state, on the RBMK reactor, the decision whether to shut down was left to the operators. A decision was made to continue the test.

In accordance with the test procedure, the steam valve to the turbo-generator No. 8 was closed, the other turbo-generator having already been shut down. At this point the reactor protection system would normally have tripped because two turbo-generators were shut down but this protection system had also been bypassed, against regulations, in order to allow the test to be carried out. It was at this moment that the reactor went prompt critical and emergency shutdown was ordered. All the reserve control and shutdown rods were driven into the core but were now unable to control the increase in reactivity. Damage to the core had already occurred and because of this some of the rods were not able to reach the low point of their travel. A number of internal rumbles were heard followed by two explosions and the reactor core started to disintegrate. The train of events following this have been described above.

12.2.6 Conclusions

12.2.6.1 *General*

The accident was caused primarily by a combination of design deficiencies and violations of rules. Using the 'time based' classification of violations described in Chapter 3, the violations leading to this accident can be described as an 'exceptional' rather than 'routine'. It is also clear from the description of the accident that many of the violations that occurred were planned since they were written into the procedure for carrying out the test. In fact it would have been impossible to complete the test without violating the normal operating rules. It is unusual therefore that the test procedure itself constituted a violation of the normal rules. Since a senior manager was present throughout the test, it must be assumed that the violations were actively condoned (if not authorized in the strict sense of the word) by management even when the plant operators raised objections to the instructions they were being given. Pre-planned violations of this nature would almost certainly fall within the rule based classification described in Chapter 3.

In Chapter 3 it was described how violations, in a similar way to errors, can have a systemic cause which, if identified and corrected can help reduce their frequency. However, because violations are by definition intentional, it was also shown in Chapter 3 that there is invariably an associated motivation. Table 3.1 in Chapter 3 listed some of these possible causes and motivations although the list was not meant to be comprehensive. Although the design deficiencies were a major factor in loss of control of the reactor, they were not themselves the systemic cause of the violations. The prime motivation for violating the rules was an almost total fixation on completing the test procedure.

As management had approved the procedure and a senior manager was present, the systemic cause of the violations must be a failure of the management systems that allowed such a test to take place.

Another feature of this accident was the progressive development of a group mindset that seemed to accept that violations were acceptable in the circumstances of the test. Once this mindset was established, it was almost inevitable that more violations would occur, whether or not they were written into the test procedure. An example of a violation that was not part of the test procedure, but became acceptable in the context of the experiment was the adjustments made to the circulating and feed water flows. These adjustments made on the spur of the moment had serious consequences for the reactivity of the core. The effects of the changes would have been understood by experienced operators who would have prevented their occurrence in normal operation. However, for the duration of the test normal rationality seems to have been suspended.

Another deviation from accepted practice occurred when the print-out for the 'reactivity reserve' was studied. The reactor should have been scrammed without hesitation, but instead the test was continued because of the mindset that prevailed. Details of the conversations in the control room are not available, but it seems clear that by the later stages of the test the plant operators had become totally compliant to the instructions being issued. This is by no means the first accident (nuclear or otherwise) where a group mindset has driven a series of actions in the face of evidence which has overwhelmingly shown it to be inappropriate.

Following the Chernobyl accident there was much debate about whether a similar accident could happen in the UK. A report by the UK Atomic Energy Authority into the accident (Gittus et al., 1988) explores this in detail. Predictably it concludes that it could not happen in the UK, but the report concentrates more on the design features of UK reactors than on the possibility of violations. This may be justified since a nuclear reactor with a positive void coefficient would never be licensed to operate in the UK. Furthermore, UK plant operators would not be permitted unrestricted access to disable safety features at will. A licensing condition of UK reactors is a design requirement that reactor shutdown systems operate fully automatically for a period of 30 minutes without operator intervention being required. This effectively locks the operating staff out of the system for the crucial stages of a reactor scram. Another feature of some nuclear power operations is the concept of the 'third man', an independent safety engineer in the control room whose main duty is to oversee the actions of the operating team. He is not part of that team and reports to a higher level in the organization than day-to-day reactor management. He is able to offer advice and has powers to veto actions that violate rules or are considered inadvisable (although it is suspected this is more than a simple veto). One of the purposes of this arrangement is to prevent the development of a group mindset in an emergency situation which could drive the response along an inappropriate path.

12.2.6.2 Human 'error' analysis

A human 'error' analysis is shown in Table 12.1. However, since it is not of errors but violations, it follows the analytical approach defined in Chapter 3, Section 3.3.5.

Table 12.1 Human 'error' analysis of the Chernobyl accident

	Systemic cause	Motivation to violate rules	Possible controls
Direct cause	Inadequate management control of violations	A series of rule based violations were written into an experimental test procedure	More rigorous management oversight of reactor operations
Root cause	Inherently unsafe design of the RBMK reactor ('positive void coefficient' and 'positive scram effect')	None – operators were aware of the dangers at operating at low power levels (but were overruled when they objected to restoring reactor power after a low power transient) and this was covered in procedures	Impossible to change operating characteristics of the reactor Better enforcement of procedures to limit operation at dangerously low power levels Procedures are always less satisfactory than an inherently safe design
Contributory causes	Unrestricted operator access to safety protection systems Decision on whether to trip the reactor at low 'reactivity reserve' was left to the operators	Convenience, avoidance of 'unnecessary' trips – safety systems could be bypassed at will allowing normal operating rules to be ignored	Clearly define consequences of violating rules Improved training
	Group mindset	Group shared beliefs promoted an acceptability that violations were a necessary part of completing the test	Presence of an independent safety engineer in the control room with access to higher management to provide advice and with power to veto or delay suspect activities

12.3 The Airbus A320 crash at Mulhouse

12.3.1 Introduction

This accident was one of a series that occurred following the introduction of a new type of 'glass cockpit' and computerized flight management system (FMS) into Airbus aircraft in the 1980s. Later, the cause of these accidents came to be known as an 'automation surprise', a scenario arising from the complexity of the computer control system coupled with a high cockpit workload. The FMS computer can be configured in various modes depending on the stage of the flight. The problem arises when an inappropriate mode is selected, for instance setting a full throttle go-around mode when the plane is about to land. Due to the many demands upon the pilot, it is easy for him to lose awareness of the current operating mode of the flight control system. When this is coupled with a lack of understanding of what each of the many computer modes is supposed to deliver in terms of aircraft performance and handling then all the precursors for an accident are in place.

The pilots involved in the Airbus A320 accident at Mulhouse found themselves in an 'automation surprise' situation. The direct cause of the accident was not a human error but a violation of fundamental flying rules strongly associated with the complexity of the computerized flight control system. It demonstrates how pilots can become overconfident in the technology to the degree that they believe the computer is sophisticated enough to compensate for errors they may make.

12.3.2 The A320 Airbus

12.3.2.1 *The 'Airbus' company*

'Airbus' is a European consortium of French, German, Spanish and U.K companies founded in 1970 to enable European aircraft manufacturers to compete effectively with large US companies such as the Boeing Aircraft Corporation. The company is based in Toulouse, France, and employs around 46,000 people worldwide. In 2001, Airbus became a single integrated company comprising the Franco–German European Aeronautic Defence and Space Company (EADS) and the UK's British Aerospace (BAe) Systems with holdings of 80 per cent and 20 per cent of the stock, respectively. Airbus manufactures a wide range of aircraft types varying from the short range A320 twin jet to wide bodied long range versions such as the A330/340 series. All Airbus aircraft use fly-by-wire controls and a 'glass cockpit' with sophisticated computer controlled FMS as described below.

12.3.2.2 *The glass cockpit*

The A320 Airbus aircraft is a single aisle commercial jet carrying 115–180 passengers and is powered by two wing mounted CFM 56-5 engines. The aircraft type was first launched in 1984 and incorporated a number of new technologies including the concept of 'fly-by-wire' controls that previously had only been used on military aircraft. The principle of fly-by-wire involves the use of computers to manage the flight control system and translate the pilot's commands into electrical impulses. These signals are delivered along wires to the hydraulic units operating the primary flight control surfaces (ailerons, elevators and rudder, etc.) replacing the conventional mechanical linkages. One advantage is a reduction in weight of the aircraft and thus improved fuel consumption. However, the main advantage claimed for the fly-by-wire system is one of safety, since an on-board computer monitors the pilot's actions to ensure the aircraft is always flown within the permitted flight envelope. This facilitates aircraft handling in an emergency avoiding conditions such as incipient stall, over-speed or over-stressing the airframe. One of the features of fly-by-wire technology in the A320 aircraft is that the conventional control column or yoke is replaced by a side-stick controller (similar to a computer joy-stick) ahead of each outboard arm rest. These allow an unrestricted view of the instrument panel and provision of a slide-out working desk in front of the pilot. Although the aircraft can be flown manually using the side-stick in the normal way, the computer control system is still operative and monitoring the actions of the pilot. If the pilot's control commands conflict with the computer's pre-programmed flight limiting parameters, then the signals to the flight controls are inhibited.

The cockpit display also deviates considerably from the conventional layout. The panel no longer comprises the traditional set of analogue electro-mechanical instruments but is modelled in the form of a 'glass cockpit' composed of six cathode ray tube (CRT) displays. In front of each pilot are two screens comprising the electronic flight instrument system (EFIS) which consists of:

- primary flight display (PFD), showing information such as aircraft speed, altitude, horizontal situation indicator and heading as well as the status of the FMS computer,
- navigation display (ND) screens showing aircraft position and course data presented in a number of alternative configurations depending on the stage of the flight.

Two screens common to each pilot are situated in the centre of the panel and show the electronic centralized aircraft monitor (ECAM) displaying engine parameters and alarms as well as monitoring system conditions throughout the aircraft via a set of sensors. On the central control pedestal are located the multipurpose control and display units (MCDUs) which are used to access the FMS via key pads for input of data which is displayed on a screen. These keys are used to enter the desired flight parameters (speed, altitude, heading, flight plan coordinates, etc.) to which the computer controlled flight system will respond. They also allow selection of autopilot and FMS operating mode, input of navigational data, fuel calculation and system maintenance data in the air and on the ground.

In case of computer failure, traditional electro-mechanical instruments are located on the centre panel to measure the essential flight parameters of altitude, air speed, horizontal attitude and compass heading. In an emergency the pilot will be able to fly the aircraft manually on these instruments to the nearest airport to effect a safe landing.

12.3.3 The accident at Mulhouse

12.3.3.1 *Preparation for the flight*
12.3.3.1.1 The demonstration flight

Mulhouse is a small French provincial town in the upper Rhine Valley situated close to the German and Swiss borders. On 28 June 1988 preparations were underway for a small air show to be held at Habsheim Airfield on the outskirts of the town. Since the main runway at the airport was only 1000 metres long and unsuitable for large jets, taking off and landing would be limited to the smaller general aviation types. However, a major event planned for that day was to be a demonstration flypast by one of the first new A320 Airbus aircraft recently delivered to Air France. The plan involved the A320 taking off from the main airport for the region, Basle-Mulhouse, situated some 29 kilometres to the south and making two low passes over the main runway at Habsheim.

The A320 aircraft to be used for the demonstration, the newest to be delivered to Air France, arrived at Basle-Mulhouse from Paris on the morning of the flight having only been delivered from the manufacturer a few days before. For the flypast the aircraft was to be piloted by Captain Michel Asseline, the Senior Management Pilot in charge of the A320 training programme and the co-pilot was Captain Pierre Mazieres, another experienced pilot and training captain. Details of the requirements for the demonstration had only been received at Air France offices in Paris during the week before the air show and as a result those details were only available to the pilots of the A320 on the morning of the air show. Neither of the pilots had ever flown from or even visited Habsheim Airfield and because of their busy schedule no time had been made available for them to carry out any sort of reconnaissance of the airfield and its surroundings.

12.3.3.1.2 The flight plan

The plan for the day, prepared by Air France, included a press reception at Basle-Mulhouse after which 130 invited passengers and sightseers would board the A320 for the demonstration flypast at Habsheim. This was to comprise a low-level approach to Runway 02 when the gear would be lowered, flaps extended and the aircraft brought down to a height of 100 feet above ground. The intention was to make a spectacular low-speed pass over the runway at a high 'angle-of-attack' (AOA) before accelerating into a climbing turn to a safe altitude, principally to demonstrate the manoeuvrability of the aircraft. It would then return for a second high-speed pass over the same runway but on the opposite heading. The demonstration flypast was to be followed, dependent on weather conditions, by a sightseeing flight south to Mont Blanc for the benefit of the passengers, before returning to disembark them at Basle-Mulhouse after which the aircraft would return to Paris.

12.3.3.1.3 The 'alpha floor' system

The AOA of an aircraft wing is the angle that it makes relative to the wind and is a fundamental parameter of flight. Clearly a wing can be angled anywhere from zero (edge-on) to 90 degrees (flat-side on) but is only able to generate lift within a very narrow angle to the wind. Most wings stall at an AOA greater than 15 degrees because the airflow across the upper surface of the wing changes from laminar to turbulent inhibiting the ability to generate lift. The AOA at which a stall will occur depends on the airspeed; the lower the airspeed, the lower the AOA at which stall will occur. The angle-of-bank (or turn) of an aircraft also has an influence on the stall speed which reduces as the bank angle is increased. In order to prevent the application of an excessively high AOA, the A320 is fitted with 'alpha floor' protection against stall and windshear accidents (when the angle of the wind changes rather than the angle of the wing). The 'alpha floor' system operates if the AOA exceeds a safe angle for the airspeed. When this occurs, the on-board flight control system automatically applies maximum take off/go-around (TOGA) thrust to restore airspeed to a safe level.

The system works in conjunction with the 'stick-shaker', a warning device fitted to larger aircraft to warn the pilot of an incipient stall condition. In older aircraft, the control yoke or stick was directly connected to the moveable control surfaces on the wings (ailerons and elevators). The onset of a stall could be recognized by the shaking of the stick as the turbulent airflow over the wing began to vibrate the control surfaces, the mechanical linkages transmitting this vibration back to the stick. On modern aircraft where the control surfaces are not connected directly to the stick, the 'stick-shaker' simulates the transmission of this vibration. Thus it simulates the 'feel' of a small aircraft that is lacking in larger aircraft. When the 'stick-shaker' operates the stick is also pushed forward hydraulically depressing the elevators to drop the nose of the aircraft thereby reducing the AOA.

In order to undertake the high AOA approach for the Habsheim flypast, it was necessary for the pilot to deactivate the 'alpha floor' protection, otherwise the aircraft would have automatically started to climb away before the low-level pass could be completed. The problem with large aircraft such as the A320 is that at low speeds and high AOAs they do not accelerate and climb very well. Even though the throttle setting is increased to maximum, it takes some time for the engines to spool up to full revolutions and then extra time before the increased power from the engines has an effect on the speed of the aircraft. This was a major factor in the accident at Mulhouse.

12.3.3.2 *The accident*

The aircraft took off from Basle-Mulhouse at 12.41 hours with 131 passengers on board on a heading of 155 degrees. Shortly after take-off, a right turn was made to put the aircraft on a heading of 010 degrees in a northerly direction towards Habsheim Airfield. Since the airfield was only a few minutes flight away, the aircraft was cleared to climb to the relatively low altitude of 2000 feet at an airspeed of 200 knots. From this height, both pilots prepared themselves for a visual sighting of the airfield. This finally came into view at about 12.43 hours. The throttle was reduced to flight idle and the aircraft commenced its descent in preparation for the low-level flypast. The airspeed gradually started to fall. By 12.45 hours the airspeed had fallen to 155 knots and the altitude was down to 200 feet. It then became apparent that the crowd line in the air show was not lined up along Runway 02 as expected but along Runway 34, one of the grass strips that was aligned in a more north-westerly direction. This required a last-minute banking movement to line the aircraft up with Runway 34. By this time the airspeed had decreased to 141 knots and the altitude had fallen to below 90 feet. A few seconds later, with the airspeed now at 131 knots and the altitude at 40 feet and still falling, the captain pulled back on the stick to raise the nose of the aircraft to an AOA of 6–7 degrees as the aircraft passed over the threshold of Runway 34. The engines were still set at flight idle or 29 per cent N_1 (per cent of full revolutions). The aircraft continued to descend, reaching its lowest altitude of 30 feet, far lower than the intended altitude for the flypast. At this point the co-pilot called out that there were some pylons and a tree line some distance beyond the end of Runway 34, clearly below the aircraft's current height (actually the trees were of average height 40 feet at a distance of about 200 metres away). The captain immediately applied full TOGA power to both engines in an attempt to gain sufficient height to bring the aircraft above the level of the treetops.

Unfortunately, the re-application of power came far too late to increase lift sufficiently to reverse the descent. It took approximately 6½ seconds for the engines to spool up from flight idle to about 87 per cent N_1 having little or no effect on the aircraft height. Due to the high AOA the tail of the aircraft struck the treetops, further increasing drag and negating the effect of increased engine power. As the engines ingested debris from the trees, they lost power altogether. As the aircraft descended into the forest the starboard wing impacted the trees and was torn away causing the fuel to empty and ignite in a fire-ball. The aircraft cut a long swathe through the forest before coming to a halt allowing passengers to exit from the rear and forward passenger doors on the port side of the aircraft. All the passengers and cabin crew except three were able to evacuate safely in spite of being impeded by tree branches and vegetation. Tragically, of the three passengers who failed to evacuate the aircraft, two were children and the third was a passenger who went back into the fuselage to rescue them but was overcome by smoke. A medical team and fire appliances from the air show arrived within a few minutes but their access was badly hampered by the fire and dense vegetation so that the aircraft was almost completely destroyed by the fire except for the tail section.

12.3.3.3 *The causes of the accident*
12.3.3.3.1 The direct cause
The direct cause of this accident was the decision by the pilots to carry out a badly planned and executed manoeuvre at low altitude from which recovery would be extremely difficult if anything went wrong. Neither pilot was qualified or experienced in air show demonstrations and carrying out such a difficult manoeuvre in a large commercial aircraft required skills that were beyond their capacity. Prior to the manoeuvre, the aircraft was not properly configured due to a shortage of time and was badly positioned on its approach for the flypast. The direct causes are therefore attributable to violations of best practice falling into two distinct categories, lack of planning and inadequate execution. In this accident the root causes, which are discussed in more detail below, can be discovered by examining the motivation of the pilots to violate best practice.

Lack of planning
The demonstration flypast was planned at short notice and neither of the pilots were familiar with Habsheim Airfield nor the surrounding terrain. It came as a complete surprise therefore when the co-pilot realized that there was a forest at the end of Runway 34. From the air the forest would have appeared as a patch of darker green in the surrounding landscape. There would have been no visual appreciation of its nature or its height above ground until the aircraft had descended to a very low altitude. For this reason, charts are issued by the civil aviation authorities for most airports showing the main approach and departure routes and indicating the position and height of any obstacles in the near vicinity. The flight dossier for the demonstration flight was only received on the morning of the demonstration. Lack of time prevented either pilot attending the briefing by the organizers at Habsheim on the morning of the show. This would have meant taking the aircraft out of regular service for even longer than the time allocated for the demonstration flight. Although a map of the airfield was included in the dossier, it was a black-and-white photocopy of a coloured original and the symbology was not clear. The time pressure partly explains how two senior training pilots with Air France violated the normal requirement for thorough preparation and flight planning before making an approach to an unfamiliar airport where they were intending to carry out a previously unpractised manoeuvre.

Inadequate execution
Routes flown by large commercial aircraft in and out of major airports follow standard arrival routes (STARs) and standard instrument departure routes (SIDs). These routes are indicated on aeronautical charts carried in the cockpit. A STAR for an airport commences at a predetermined navigational fix such as a Very High Frequency Omni-directional

Radio (VOR) Beacon at a specific altitude and heading (see Chapter 11 for an explanation of aircraft navigational procedures). The aircraft is often positioned some 10 or 20 miles from the runway threshold before the approach is commenced, allowing sufficient time for the aircraft to be put into a landing configuration. Most large airports also use an instrument landing system (ILS) for the main runways comprising a directional radio beacon transmitting a signal with vertical and horizontal components. The vertical component is known as the glide slope indicator. It may be connected to the autopilot allowing the aircraft to fly the approach automatically so that the aircraft is always at the correct altitude with respect to distance from the runway. It will then arrive at the threshold at the optimum landing height. The horizontal component is known as the localizer and can also be connected to the autopilot. This ensures that the correct heading is maintained so that the aircraft will arrive at the threshold with its nose wheel on the runway centreline. Pilots are trained in the use of these landing aids.

This principle of setting up an approach to a runway gives pilots time to configure the aircraft for landing and applies to light aircraft as well as heavy commercial types. It is part of the basic training of all aviators. Perhaps the only exception would be emergency situations when an aircraft is required to land on the nearest runway as quickly as possible. Even in these cases, air-traffic control (ATC) would use radar to vector the aircraft on to the approach for the runway by radio instructions to the pilot. The pilots on the short flight to Habsheim had made no such preparation for an approach and were flying visually without the assistance of air-traffic controllers. The aircraft was badly positioned for the manoeuvre right from the start. It was flying too low and too close to the airport for a controlled approach to be set up and there was insufficient time to configure the aircraft correctly in terms of height, speed and flap settings, etc.

At the inquiry into the accident (Investigation Commission, 1989) it was questioned why the aircraft had descended so far below the planned altitude for the flypast. Once the throttles had been set to flight idle, the records from the black box showed that the aircraft lost height consistently at a rate greater than was necessary to achieve the planned flypast height. It was probable that the pilot's attention was taken up with locating the position of the airfield and the runway instead of monitoring the height and speed of the aircraft.

Another distraction arose when they realized that the crowd line at the air show was stretched out along a grass runway and not the main concrete runway as expected. This required a last minute change in heading at a time when they should have been lining up the aircraft on the correct approach path. A last-minute attempt was made to regain speed and altitude but at a height of only 30 feet above the ground recovery was impossible.

Perhaps the most important factor in this accident was the pilots' over-confidence in the aircraft's computer controlled fly-by-wire system which, they seemed to believe, made the aircraft immune to pilot error. In reality though, the aircraft was no different from any other in that it needed to conform with the physical laws of flight to remain airborne. This over-confidence in automation is considered to be the root cause of the accident and is discussed in more detail below.

12.3.3.3.2 The root cause
The root cause of this accident was an over-reliance on automation. Due to this the pilots were motivated to violate a number of fundamental flying principles. It appears that the

sophisticated levels of automation incorporated into the fly-by-wire A320 aircraft, invited the pilots to carry out manoeuvres that would be considered dangerous in any other aircraft. In this respect the violations were knowledge based in the sense that they were not what 'a reasonable or experienced person (in this case a qualified pilot) might be expected to do' in the circumstances (see Section 3.3.4.3). Their confidence in the systems was so great that it displaced elements of basic training and flying experience.

A major misconception might have arisen from the training given to A320 pilots on the new aircraft. This training had recently been delivered to Air France pilots by the two training captains on this flight. It placed great emphasis on the protection provided by the aircraft's computerized flight systems against exceeding the limits of the permitted flight envelope. What the training did not make clear was that these limits on aircraft performance still existed, and that the normal laws of aeronautics did not cease to apply simply because a computer was monitoring the actions of the pilot. If air speed falls to a level where the wing provides insufficient lift then as with any other aircraft it will lose height and eventually stall. This is what appears to have happened at Habsheim. As the aircraft struck the trees the air speed had fallen to 112 knots, well below the stall speed. At this point the aircraft ceased flying.

The computer system allows the limits of the permitted flight envelope to be approached but then blocks any pilot input to the controls that would exceed these limits. However, as the limits of the envelope are approached, the performance reserve of the aircraft becomes significantly eroded and although straight and level flight can be maintained, recovery from extreme manoeuvres (such as a high AOA or a steep turn, for instance) may not be possible. On the approach to Habsheim, one of the more important protection devices, the alpha floor protection, was deactivated. Normally this would have applied full TOGA power automatically before the aircraft had reached a dangerously high AOA, allowing recovery to take place. Indeed recovery might still have been possible if the aircraft had possessed sufficient altitude but with the altitude below 100 feet and a descent rate of 600 feet/minute recovery was impossible in the time before the aircraft struck the trees. If it had not struck the trees it would probably have flown into the ground. During the inquiry, Captain Michel Asseline insisted that the engines had not delivered the expected performance, and that when he applied full throttle, there was a delayed response. However, data from the flight recorder proved that this was not the case and the performance of the engines after the throttles were advanced was entirely within specification.

12.3.3.3.3 Contributory causes
Cockpit resource management
Conversations between the captain and co-pilot before and during the flight indicate that the co-pilot had some misgivings about the manoeuvres to be carried out. As the aircraft was taking off the captain described how the flight would be conducted mentioning that the low-level flypast would be made at 100 feet followed by a climbing turn when the co-pilot would apply full power. When the co-pilot expressed some concern about this, the captain assured him that he had 'done it about twenty times before'. During the last few seconds of the descent to Habsheim and at about 40 feet above the ground, the co-pilot suddenly warned the captain of the pylons that were ahead. The captain replied

'yes, do not worry' and continued the descent. The problem of a subordinate being over-ruled by a superior in the cockpit was mentioned in Chapter 11 in connection with the Boeing 737 accident at Kegworth. Many airlines have instituted training in cockpit resource management (CRM) techniques in order to improve communications between aircrew, cabin crew and in some cases passengers.

Undue company pressure

There was pressure from Airbus and Air France to demonstrate the capabilities of the A320 control system particularly in handling a low-speed high AOA flypast followed by a dramatic climb and accelerating turn. It is probable that this sort of pressure may have helped overcome any initial reservations the pilots might have had in conducting an unfamiliar manoeuvre. French air safety regulations actually banned demonstration flights of this type below a height of 170 feet above ground level, although these regula-tions had been breached on a number of previous occasions.

A 'holiday' atmosphere

The flight was conducted with 130 invited passengers on board, some of whom were senior executives. This seemed to induce a 'holiday' atmosphere with considerable excitement and anticipation. It is probable that this sort of atmosphere worked against the natural conservatism of the pilots and may have encouraged a breach of flying rules that would have prevailed in a normal situation. They would have been keen to impress their guests by showing off the aircraft's performance in a spectacular manoeuvre.

12.3.3.4 'Automation surprise'

The accident at Habsheim was one of a group of accidents which occurred over a period of about 5 years following the introduction of Airbus aircraft using the same type of fly-by-wire glass cockpit. A common factor in all these accidents was the variety of flight management mode settings which were selectable ranging from manual to fully auto-matic control. Different modes will be selected for each phase of the flight, for instance, for take off, climb, level flight, descent, approach, landing and go-around. One of the problems associated with these cockpits has been a failure by pilots to recognize or remember which mode is currently selected. Adding to this confusion is uncertainty about what the selected mode will deliver in terms of the degree of flight automation and the expected performance of the aircraft. The pilot must always be aware of what can be left to the computer and what has to be carried out manually. He has to recognize at all times the relative allocation of function between human and machine. This may be quite difficult to assess in the busy cockpit environment when an aircraft is flying a complex approach through crowded airspace following directions from ATC. These problems have led to the situation of 'automation surprise'. It often occurs when a mode is selected for an earlier phase of the flight and not changed when a new phase is entered. Alternatively an error may be made in selecting the mode for the next phase. Some-times the error is not discovered until the aircraft fails to respond or responds in an unex-pected way warning the pilot that an inappropriate mode is selected. By this time it may be too late to make a recovery. The following are a small selection of A300/320 series aircraft accidents involving automation surprises.

12.3.3.4.1 Bangalore, 1990

On 14 February 1990, an almost new Airbus A320 aircraft of Indian Airlines on a flight from Bombay crashed on a manual approach to Bangalore airport. Although Bangalore airport is not equipped with an instrument landing system the aircraft should have been able to make a safe visual approach under manual control. However, about a kilometre short of the runway the aircraft adopted a high AOA and began to fall below the glide slope impacting with the ground about 700 metre short of the threshold. This was very similar to the Habsheim accident, since full throttle was only applied when the aircraft was 140 feet above the ground. As at Habsheim, maximum power application was insufficient to arrest the high rate of descent. As a result 92 of 146 people on board died in the accident and the fire that followed. This aircraft was being flown by the co-pilot monitored by a senior training pilot who himself only had 68 hours experience on the aircraft type.

The inquiry into the accident found that the FMS had been set for the descent in the open descent idle mode (ODIM) where the throttle is held continuously at flight idle. This was in contravention of the airline's operating manual. As the aircraft lost height due to low engine power the computerized flight system lifted the nose and attempted to increase power to the TOGA setting as if for a go-around. However this was prevented because the computer had set the throttles at ODIM, as selected by the pilot who, had either forgotten or was not aware of the mode setting during the final critical moments of the descent. The accident resulted in Indian Airlines placing a temporary ban on further deliveries of A300 type aircraft until the problems were properly understood and resolved.

12.3.3.4.2 Nagoya, Japan, 1990

On 26 April 1994, an Airbus A300 aircraft of China Airlines on a flight from Taipei crashed when making an ILS approach to Nagoya airport under manual control. During the approach phase the co-pilot inadvertently activated the go-around button switching the mode of the FMS causing TOGA (full) throttle to be selected. This caused the aircraft to climb above the glide slope and in response to this the pilot pushed the stick forward at the same time engaging the autopilot to capture the ILS localizer beam. This resulted in the elevators moving to put the aircraft into a high AOA. The aircraft now attempted to climb out at a very steep angle and at this point the captain decided to take control and make a go-around for another approach. However, the aircraft stalled, control was lost and it crashed close to a taxiway within the airport perimeter. The aircraft burst into flames and was destroyed resulting in the deaths of 264 persons on board. Throughout the descent, both pilots had believed that they could land the aircraft manually using the side-stick, not realizing that the engagement of TOGA thrust made this impossible. They failed to recognize the conflicts between their manual input to the aircraft controls and the response of the aircraft's computerized control system.

12.3.3.4.3 Strasbourg, France, 1992

On 20 January 1992, an Airbus A320 passenger jet of Air Inter flying from Lyon to Strasbourg crashed in darkness into mountains in eastern France. This was a 'controlled flight into terrain' accident since both pilots had been completely unaware that a crash was imminent until they received a low altitude alert at 200 feet, by which time it was too late to avoid the high ground. The cause of the crash was a mistaken and inappropriate input to the computerized FMS. An attempt to land at Strasbourg had already been made and the aircraft was in the process of a go-around being vectored on to the approach path by ATC. As a result, both captain and co-pilot had a high workload in trying to ensure that they were on the correct flight path at the right descent rate at the same time as they were trying to complete the pre-descent checklists in the short time period before landing.

A number of major inputs needed to be entered into the computerized FMS. The rate of descent and the angle of descent were both entered using the same dial but the appropriate mode needed to be set. The co-pilot punched in a value of '33' believing that he was programming the computer to give a 3.3 degrees glide slope command. However, the mode was actually set at 'vertical speed' mode so what he had entered was a descent rate of 3300 feet/minute. While an entry of 3.3 degrees would have given the required gentle descent to Strasbourg airport, a vertical descent speed of 3300 feet/minute resulted in a rapid descent to below the peak of mountains located between the aircraft and the airport. This incorrect mode selection led to an accident causing the loss of all on board except eight passengers and one flight attendant.

All the above accidents involved human error rather than violations, but provide an illustration of how a new design of aircraft using sophisticated computerized flight control was beyond the capabilities of even experienced pilots who had undergone conversion training. It shows how 'paper' knowledge of the various systems and interactions is inadequate in the real situation of handling an aircraft with a high cockpit workload. Since these accidents, major changes have been made to the technology and to the way training is delivered, particularly in ensuring pilots are aware of their own vulnerability to 'automation surprises'.

12.3.4 Conclusions

12.3.4.1 *General*

As described in Section 12.3.3.4, this accident was one of a number which occurred following the introduction of the computerized glass cockpit into various types of Airbus aircraft. The accident involved a knowledge based violation of fundamental flying rules rather than human error. However, as described in Chapter 3, Section 3.3.4.3, knowledge based and skill based violations are sometimes difficult to differentiate from errors. What needs to be considered is the degree to which the abnormal action was intentional. In this case it was intentional but motivated by an over-confidence in the ability of the aircraft's computer system to compensate for a series of pilot inputs that would not have been acceptable on a normal aircraft. The pilots were fully aware of the limitations that would have prevented a normal aircraft undertaking these manoeuvres safely. It was in full knowledge of these limitations that they decided to go ahead with the manoeuvre and for this reason the misjudgements that were made are classified as violations rather than errors. At the same time it needs to be recognized that by the time they were over

Table 12.2 Human 'error' analysis of the Mulhouse A320 accident

	Systemic cause	Motivation to violate rules	Possible controls
Direct cause	A badly planned and executed manoeuvre at low altitude from which recovery would be extremely difficult if anything went wrong	Over-confidence (see below), shortage of time to complete demonstration	Improved management control of non-standard flying operations
Root cause	Over-reliance on automation (aircraft computerized FMS)	A seeming belief that the fundamental principles of flight could be violated with impunity	Training to include a more realistic appraisal of the aircraft's performance limitations
Contributory causes	Co-pilot failed to transmit his concerns sufficiently strongly to the captain (a more senior figure)	A desire not to appear unwilling to participate in the demonstration	Cockpit resource management training
	Effects of peer pressure	It was easier to concur with the plan since both pilots were senior airline captains	Cockpit resource management training
	Undue pressure to demonstrate the aircraft's performance	Pressure from airline and plane manufacturer to show off aircraft handling characteristics	A culture which avoids conflicts between commercial considerations and safety
	A 'holiday' atmosphere on board the aircraft	Public relations – a desire to please a select group of invited passengers	As above

the aerodrome their options were severely limited by inadequate flight planning and a failure to be properly briefed about local conditions.

12.3.4.2 *Human 'error' analysis*

The human 'error' analysis in Table 12.2 is not of errors but violations. It therefore follows the analytical approach defined in Chapter 3, Section 3.3.5. It is a summary of the salient points in Section 12.3.3.3.

References

Gittus, J.H. *et al.* (1988). *The Chernobyl Accident and Its Consequences*, 2nd edition, London: UKAEA.

International Atomic Energy Agency (1997). International Initiative for the Chernobyl Sarcophagus, htttp://www.iaea.org/worldatom/About/GC/GC41/resolutions/gc41res21. html.

Investigation Commission (1989). *Inquiry into the Accident on June 26th 1998 at Mulhouse-Habsheim to the Airbus A320 registered F-GFKC. Final Report*, Ministry of Planning, Housing, Transport and Maritime Affairs. Official Translation from the French by the Bureau d'Enquetes Accidents (BEA).

Incident response errors

13.1 Introduction

Incident response activities are usually carried out to limit or mitigate the consequences of an adverse event or put into effect recovery operations to normalize a situation. It is immaterial to incident response whether or not the incident was caused by a human error since the activities which follow are determined not by the cause but by the consequence. The types of errors involved in incident response are no different from those described in Part I and dealt with in earlier case studies.

The adverse conditions under which incident response tasks are carried out sets them apart from normal tasks. The situation following an incident is often classed as an emergency, depending on the severity of the consequences. Emergencies are characterized by elements of surprise and uncertainty so that the tasks which follow tend not to be well practised and may involve high stress levels. They are often carried out under time pressure with high workload. These are all factors which tend to increase human error probability. Incident response errors are more likely to be active than latent since the success or otherwise of the task outcome tends to be revealed immediately. However, latent errors can play a part in hindering the effectiveness of incident response due to inadequate preparedness in terms of equipment, procedures or training.

Due to the unpredictable nature of an emergency, few of the tasks are likely to be skill based, except perhaps for activities by members of emergency services who provide a skilled response based on training and experience of similar incidents. This will include familiar tasks such as operating fire and rescue equipment. What may be an abnormal situation to a person unprepared for an emergency may be a normal occurrence to a member of the emergency services. However, there is usually some element of pre-planning for incidents and emergencies so there will also be rule based tasks using emergency procedures. These procedures must be carefully prepared since accessibility and usability will be important in the adverse conditions of an emergency.

At management and supervisory level the emergency situation may not have been experienced before and the tasks are more likely to be knowledge based. Knowledge based activities involve a strong element of diagnosis and decision-making. Initially a diagnosis is made to establish what went wrong followed by the selection of a suitable plan of action based upon the diagnosis. This is followed by execution of the plan with the possibility of errors. In emergency situations there is considerable scope for 'mistakes' to occur whereby an inappropriate plan is selected or devised. The plan may be correctly executed but an unsuccessful outcome may reveal it was based on a faulty diagnosis. As discussed earlier, the probability of errors in knowledge based activities is much higher than in skill or rule based activities.

Both case studies in this chapter involve the hazards of fire and smoke, involving an aircraft and a railway tunnel, respectively, environments where people may become trapped. The hazard of smoke in an aircraft cockpit also has the potential for loss of control. This is exactly what occurred in the Swissair flight SR111 accident over Nova Scotia in 1998. The aircraft was destroyed due to impact with the sea, but the root cause was smoke in the cockpit. The case study demonstrates the problem of carrying out emergency procedures under adverse conditions. In the Channel Tunnel fire of 1997 there was no loss of life but injuries occurred together with major economic consequence of loss of utility of the tunnel for a prolonged period with erosion of public confidence.

Of interest for these case studies is the cause of the errors that were made during the response to the incident. In both cases the designers failed to recognize the limitations on human performance in emergency situations due to adverse conditions. In addition there were a number of latent errors associated with the design of equipment, systems or procedures that were only revealed as the incident unfolded.

13.2 Fire on Swissair flight SR111

13.2.1 Introduction

The incident which led to the loss of Swissair flight SR111, a wide bodied passenger jet en route from New York to Geneva in 1998, was an electrical short circuit leading to smoke in the cockpit. Control of the aircraft was lost and it descended into the sea off Peggy's Cove, Nova Scotia with the loss of all on board. Any aircraft emergency results in high workload and stress for the flight crew. Add to this the debilitating effects of smoke in the cockpit and the probability of successfully handling the emergency is considerably reduced. In spite of this, there have been incidences of smoke in the cockpit where the flight crew were able to recover and land safely. There are two main areas of interest: the reason for the fire in the cockpit, which was the incident requiring a response, and the subsequent loss of control by the flight crew, which was the direct cause of the accident. While the cause of the fire is discussed briefly below, of most interest for this book is the underlying reasons for the loss of control. The case study demonstrates the importance of providing user-friendly systems and procedures capable of supporting incident response tasks under adverse conditions. This is the responsibility of the designer and supplier of equipment, in this case the aircraft manufacturer, as well as the operator, which is the airline. As in most of the case studies in this book, although the direct cause of the accident may be a 'human error', the root cause lies in the inadequate systems provided to support the task.

13.2.2 The accident

Swissair flight SR111 with 229 passengers on board left JF Kennedy airport, New York at about 20.20 hours local time on 3 September 1998 bound for Geneva, Switzerland. The aircraft was a MD-11 a modern wide-body tri-jet aircraft built by McDonnell Douglas Corporation. The aircraft, which is no longer manufactured, is an upgraded version of the much older DC10 aircraft which had been in service on transcontinental routes since the 1970s. About an hour after departure the flight was approaching Nova Scotia at 33,000 feet when the aircrew detected an unusual smell in the cockpit. A few minutes later, traces of smoke began to appear and the crew transmitted the international urgency signal 'Pan Pan' to the Monkton (Novia Scotia) high-level air-traffic controller. Shortly after this conditions in the cockpit began to get worse. The aircraft's flight data recorder, recovered after the accident revealed that by this time the autopilot had been disconnected, followed by a rapid series of failures of various aircraft systems. Transcripts of the voice communications with air-traffic control (ATC) show that the pilot and co-pilot had donned oxygen masks.

Within 15 minutes the crew had upgraded the incident to an emergency with a request to be vectored to the nearest airport. The flight was cleared to descend to a lower altitude and was diverted to Halifax International. At 21.20 hours the flight was handed over to the Moncton centre, which vectored the aircraft for an approach to Halifax Runway 06. It was requested that the aircraft make a turn to the south, and enter a series of circling manoeuvres to lose height before dumping fuel over the sea in order to allow the approach and landing to be made. The reason for this was that, in common with many large commercial aircraft, the MD-11 has a maximum landing weight. This weight is exceeded at take-off due to the fuel on board for a transcontinental journey. Fuel must be dumped if a landing is required shortly after departure. Within a few minutes the crew again contacted Moncton centre to report that they had commenced dumping fuel and needed to make an immediate approach to the runway. The aircraft was now flying at between 5000 and 12,000 feet over the sea. Monkton centre requested the aircraft to indicate when the fuel dump was complete. After this no further voice communication with the aircraft was received and the radar image disappeared from view when the aircraft was about 35 nautical miles from the airport.

The accident was investigated by the Transportation Safety Board of Canada (TSBC), necessitating the recovery of wreckage from the seabed, an exercise which lasted 2 years. During this time, the black box recorder was recovered and provided important information about the state of the aircraft systems during the emergency. The final report of the TSBC was published on 27 March 2003 (the Transportation Safety Board of Canada, 2003) on completion of extensive technical investigation and analysis work extending over a period of nearly 5 years. The causes of the fire in the cockpit were identified in the early stages of the investigation and as a result immediate recommendations were made regarding the insulation material used in the wiring of the cockpits of these and similar aircraft. The causes of the fire are briefly discussed below.

13.2.3 The causes

13.2.3.1 *The cause of the fire*

In-flight fire is one of the most serious situations to occur on an aircraft and has caused a number of fatal accidents around the world. The greatest hazard arises from loss of control of the aircraft due to the flight crew being overcome by smoke and loss of visibility in the cockpit preventing the aircraft being flown or landed safely. In order to overcome this problem the flight crew will normally don oxygen masks provided for loss of cabin pressurization. Automatic fire fighting equipment on aircraft is only fitted in 'designated fire zones'. These are normally defined as areas inaccessible for fire fighting and which contain combustible materials and a source of ignition. These areas would include for instance power plants, auxiliary power units, cargo areas and toilets. In general the materials used in the cockpit are supposed to be fire resistant and designers do not consider the cockpit an area that would contain recognized fuel and ignition sources.

A number of cockpit fire incidents had occurred on McDonnell Douglas aircraft prior to the SR111 accident:

1. On 18 January 1990 an MD-80 cockpit became filled with smoke as a result of electrical wire insulation becoming overheated due to a power feeder cable terminal melting from intense arcing.
2. On 16 October 1993 smoke entered the cockpit of a German MD-81 as a result of an electrical short circuit behind the overhead panel. The pilots were hindered by dense smoke and unable to read emergency procedures.
3. On 8 August 2000 an in-flight fire was experienced on a US registered DC9 at take-off, causing the crew to don oxygen masks and smoke goggles. Their visibility was so severely restricted that they could not see the cockpit instruments or outside the aircraft. It was later found to be due to wiring defects behind the pilot's seat.
4. On 1 October 2000 an electrical fire occurred onboard a US registered MD-80 just after take-off caused by short circuits of a number of heavy-gauge electrical wires causing sparks and the cockpit to fill with smoke. Later the wires were found to be welded together.

Fortunately, none of these accidents resulted in an accident but their prevalence on this type of aircraft seems to indicate a problem with electrical wiring.

Investigation into the wreckage of SR111 recovered from the seabed revealed a high degree of fire damage above the ceiling on the right side of the cockpit in the forward section of the aircraft. About 20 electrical wires were found to have been subject to an arcing event leaving behind traces of melted copper. Most of the wiring on an aircraft is located behind panels and other inaccessible areas and groups of wires are bundled and bound together. This makes it difficult to inspect the condition of electrical wiring that may become damaged as a result of vibration, contamination or chafing. This can result in the breakdown of insulation and possible arcing when the wires come in contact.

The Federal Aviation Administration (FAA) of the US had previously conducted a number of tests as part of its ageing aircraft programme to examine the condition of critical wiring in older aircraft. It was found that the insulation of many wires had degraded over time leading to severe damage from vibration, cracking, delamination and arcing. An exposed wire can conduct current perfectly normally without any effect on aircraft performance until the exposed metal is shorted to earth. The effect upon aircraft performance can then be catastrophic. This is especially the case with modern aircraft using fly-by-wire systems instead of cable and pulley mechanisms to operate the main flight controls as on older aircraft. It is estimated that a modern large commercial aircraft contains about 250 kilometres of wiring. Electrical systems have been implicated in about half of all events of in-flight fire or smoke and problems with wiring account for about 10 per cent of these.

An arcing event in a modern aircraft, apart from the disruption to aircraft systems, is also a potential source of fire if inflammable materials are present. Tests carried out by the TSBC revealed deficiencies in the combustion characteristics of insulation materials used to protect the wiring and other parts of the aircraft. In particular the tests revealed that the cover material for thermal acoustic insulation blankets was coated with metallized polyethylene terephthalate (MPET) film, a material sold commercially as *Mylar*, which was able to propagate fire. The TSBC report states that in the SR111 accident 'the cover material was most likely the first material to ignite, and constituted the largest portion of the combustible materials that contributed to the propagation and intensity of the fire'. It was discovered that most of the smaller MD-80 and -90 jets built by the same manufacturer, and about 1 in 4 of the older DC-10 aircraft, contained this type of insulation material. Use of the material had been banned in US military aircraft some decades before this accident. *Mylar* insulation on the MD-11s flown by Swissair has now been replaced and recommendations made by TSBC regarding specification, testing and installation of insulation and electrical wiring. These recommendations are likely to be put into effect in the form of improved regulations concerning specification of electrical systems on new aircraft, and some modifications to existing aircraft.

13.2.3.2 *The reason for loss control*

The direct cause of the accident was loss of control of the aircraft due to incapacitation of the pilots. In order to understand how this occurred it is necessary to examine the emergency procedures for smoke or fire in the cockpit due to a suspected electrical fault.

The loss of control could have been caused by malfunctions, such as loss of instrumentation or a failure of one of the flight systems. It is unlikely that a malfunction occurred which was serious enough to prevent the aircraft being flown, since it performed 360 degrees turns to burn off fuel after the initial advisory PAN declaration. If such a serious malfunction had occurred, then it is probable that a distress or Mayday would have been made to Monkton ATC much earlier. Once the problem of smoke in the cockpit had become apparent, the pilots would have commenced a procedure for 'smoke/fumes of unknown origin' guided by a checklist issued by Swissair. Before the procedure was commenced, the pilots would have donned oxygen masks and goggles to protect them from the effects of the smoke and assist visibility and comfort. The wearing of this equipment would hamper every task that was subsequently carried out.

When the requirements of the procedure are examined, it quickly becomes apparent that the demands upon the flight crew are excessive. The procedure for following a checklist is that the first officer calls out the items on the list and the captain will then check or execute them. Since one of the likely causes of a cockpit fire is an electrical fault within the wiring, the checklist calls for various systems to be electrically isolated in a predefined order. This involves isolating various cabin electrical power breakers, pausing long enough to evaluate whether smoke or fumes have decreased. If there is a decrease, the power breaker is left isolated and the aircraft landed. The procedure also calls for a 'SMOKE ELEC/AIR Selector' to be turned clockwise, again pausing at each position long enough to check whether smoke or fumes decrease. Rotation of the selector will lead to a shutdown of essential aircraft systems in turn. The aircraft systems will be disengaged and become unusable. For instance, the autopilot and auto-throttle will disengage and problems may arise with the stick shaker that warns of an imminent stall. Some of the modes of the computerized flight management system become unavailable. Each position of the selector switch isolates a different electrical circuit which provides power to a number of separate aircraft systems. The checklist advises the pilots which systems are unavailable at each position of the selector switch. The procedure terminates with a statement 'if smoke/fumes are not eliminated, land at nearest suitable airport'.

The diagnostic process is therefore one of elimination and waiting to see whether there is a change in smoke conditions in the cockpit. At the same time as the procedure is being put into effect, the flight crew are still required to fly the aircraft, navigate to the designated airport as vectored by ATC, operate the fuel dump procedure, communicate with the ground, pass instructions to the cabin crew and make announcements over the public address (PA) system. All this is carried out while wearing oxygen mask and goggles.

The electrical isolation procedure may not clearly indicate the cause of the problem since there may not be enough time to determine whether the smoke has diminished after each circuit is isolated. Confusion about the origin of smoke may also arise from smoke being moved around by the air-conditioning system. The theory is that if the smoke dissipates, then the problem is solved and conditions should improve to allow a safe landing. The circuit that is causing the problem will then, of course, need to remain isolated. However, while the procedure is being carried out the aircraft is being flown in a degraded condition, at times manually in conditions of poor visibility, with the autopilot disconnected. During the procedure, important alarms and aircraft flight configuration warnings will be switched off at a time when they are most likely to be needed. Some of the important aircraft systems which are disabled are the multipurpose control and display units (MCDU) 1 and 2 used to access the flight management control system, problems with stick shaker for stall warning, all navigation aids and radar, pitot head heating for accurate altitude measurement, auto slat extension, flap limit selector and engine reversers for landing, landing gear aural warning, auto brakes, go around mode unavailable, fuel dump low-level, horizontal stabilizer trim switches on control column, etc. With these essential systems disabled a loss control of the aircraft is made more likely.

Modern aircraft are provided with advanced flight management systems designed to reduce the workload of the flight crew in normal conditions. In emergency conditions, these systems have an even greater utility. It seems anomalous therefore, that in the recovery from an emergency, many of the systems provided to assist the flight crew are disabled. As the emergency progresses the workload increases, the support from the aircraft systems becomes less and the chances of a successful recovery are reduced.

An alternative theory for the SR111 accident has been advanced but never proven. It is suggested that the flight crew were so preoccupied with running through the check list and shutting down electrical systems while the fuel dump was proceeding that the quantity of fuel remaining went unmonitored until the aircraft ran out of fuel and descended into the sea. One of the items disabled by following the 'smoke/fumes of unknown origin' procedure was the fuel dump low-level alarm (see list above). However, this scenario is thought unlikely since if engines had started to flame out then this would have been transmitted to ATC before the aircraft crashed.

The reason for the loss of flight SR111 is summed up succinctly by the TSBC report which states, 'the loss of primary flight displays and lack of outside visual references forced the pilots to be reliant on the standby instruments for at least some portion of the last minutes of the flight. In the deteriorating cockpit environment, the positioning and small size of these instruments would have made it difficult for the pilots to transition to their use, and to continue to maintain the proper spatial orientation of the aircraft'. The adverse conditions under which the flight crew were required to carry out their tasks must have had a major influence upon the outcome of their response. In the end, handling an emergency of this type comes down to a problem of how best to utilize limited time and attentional resources. Knowing what you can ignore is almost as important as knowing what you should pay attention to. The trial and error nature of the procedure for isolating electrical systems meant that every system in turn needed to become the focus of attention, irrespective of its likelihood of being the culprit. This seems to be an inefficient way of marshalling resources, and is further discussed in the following section.

13.2.4 Conclusion

13.2.4.1 *General*

The accident was initiated by a design fault involving the use of insulation material capable of propagating fire following an electrical short circuit. The adverse cockpit conditions then reduced the ability of the flight crew successfully to carry through the 'smoke/fumes of unknown origin' procedure and at the same time fly and land the aircraft. When the emergency procedures are specified an assessment must also be made of whether the task demands are within the reasonable capability of those required to carry them out. The assessment should include the potential for adverse performance shaping factors brought about by the emergency conditions to which the procedure is designed to respond.

The assessment ideally requires human factors specialists to assess the vulnerability to human error of the response tasks. Quite often in large projects, the usability aspects of a design are not studied until the design process is all but completed. The problem then arises that when operability aspects of equipment are found to be unsatisfactory, it becomes very expensive to change the design. By this time the system may be under construction or even worse, may have been commissioned so that the problems are not discovered until the human is introduced to the system for the first time. As in this case study, the problems may not be realized until an emergency occurs by which time it may be too late to prevent a fatal accident.

When it is too late to change the design, the problems are sometimes addressed by devising procedures to minimize the impact, on the principle that procedures are easier and less costly to change than equipment. It is very unlikely that this approach can provide an adequate solution since procedures devised to overcome the design deficiency tend to be complex and difficult to implement. It is another example of a 'design-centred approach' (whereby the human must adapt to the machine) being adopted rather than a 'user-centred approach', as discussed in Chapter 4, Section 4.1.

The use of the 'smoke/fumes of unknown origin' checklist is in line with the rule based approach generally adopted for handling aircraft emergencies. If the process of diagnosis of a fault condition were purely knowledge based, that is working from first principles, then the chances of making a correct diagnosis under conditions of stress are

Table 13.1 Human error analysis of the Swissair flight SR111 accident

Cause	Description	Type of error or other failure	Systemic cause or preventive action
Direct cause	Initiating event – electrical wiring fault which led to fire and smoke emission.	Design error	Unsuitable insulation material specification, inadequate material testing due to regulations not being sufficiently rigorous.
Root cause	Failure of flight crew to retain control of aircraft under adverse cockpit conditions.	Active errors	Systems and diagnostic procedures to deal with cockpit fire which did not take account of the cockpit human factors – in particular high workload and threat stress under debilitating conditions. Designer's lack of awareness of cockpit human factors.
Contributory causes	Flight crew were misled by smoke moved by the air-conditioning system into thinking this might be the source of the problem.	Diagnostic (knowledge based) error	Provide a more structured knowledge based diagnostic approach to identifying the most likely cause prior to selecting a rule based procedure for eliminating the problem.

extremely small. Hence the principle of making the flight crew follow a rule based procedure is a correct one and will increase the probability of success given that the task demands are reasonable. The provision of a written checklist by Swissair ensured that the diagnostic process was followed methodically in the correct order reducing the chance of omissions being made. The problems arose from the limited capability of the pilots to carry out such a demanding procedure in the adverse conditions and still fly the aircraft.

13.2.4.2 *Human factors analysis*

Table 13.1 summarizes the main human error causes lying behind the accident in terms of its direct, root and contributory causes and tabulates these causes against the different types of error that occurred.

13.3 The Channel Tunnel fire

13.3.1 Introduction

The major hazard of road and rail tunnels is the occurrence of fire with all the inherent difficulties of evacuation of people and of getting emergency services to the scene. A number of serious tunnel fires have occurred in Europe in recent years, the worst being the Mont Blanc Tunnel fire in 1999 (39 fatalities) and two tunnel disasters in Austria, the Tauer tunnel accident in 1999 (12 fatalities) and the Pfänder Tunnel accident in 1995 (3 fatalities). Five other tunnel disasters that have occurred not involving fatalities include a fire in the Norwegian Seljestad tunnel (2000), two fires in Italy in 1997 (the rail–car incident in the Exilles tunnel and the Heavy Goods Vehicle (HGV) incident in the Prapontin tunnel), a fire in the Munich Candid tunnel (1999) and the Channel Tunnel fire of November 1996 involving a fire on a HGV Shuttle train. The latter is the subject of this case study.

Although no fatalities occurred as a result of the fire, all the precursors were in place for a major disaster involving not only the occupants of the incident train, but the many thousands of passengers in trains in both Running Tunnels. The potential consequence in terms of loss of life and future utility of the rail link were immense. The case study is not only instructive in terms of understanding the reasons for the response errors, but is also an example of how a design modification made for operational and economic reasons, almost turned an incident into a disaster. Fatalities were only averted because of fortuitous circumstances which allowed the safe evacuation of HGV drivers and train crew.

13.3.2 Channel Tunnel description

On 12 February 1986 the foreign ministers of England and France signed the Franco–British Treaty to build the Channel Tunnel and awarded a concession to Eurotunnel PLC to carry out the development, financing, construction and operation.

The Channel Tunnel is a 50 kilometre long fixed rail link between England and France, 37 kilometres of which are under the English Channel. It was one of the major infrastructure projects of the late 20th century and is unique in Europe in terms of length and volume of traffic, having carried over 50 million people since it opened in 1994. The Tunnel operates a variety of train types between Folkestone in the UK and Coquelles in France.

The Channel Tunnel comprises two single track Running Tunnels each 7.6 metres in diameter with traffic flowing unidirectionally from England to France in the Running Tunnel North and from France to England in the Running Tunnel South. The track comprises continuously welded rails laid on pre-cast concrete supports embedded in the floor of the Tunnel. Ancillary equipment such as fire water mains, signalling cables and equipment are fixed to the sides of the tunnels. Traction power to all electric trains is supplied from an overhead catenary divided up into segments so that maintenance work can be carried out with the Tunnel in operation.

In between the two Running Tunnels is a smaller 4.8 metre diameter Service Tunnel with access to each of the Running Tunnels via Cross Passage Doors at 100 metre intervals. These are used for access to the Tunnel by emergency services as well as for maintenance. The Service Tunnel also acts as a safe haven to which passengers can be evacuated from trains in the Running Tunnels in the event of an emergency. The integrity of the Service Tunnel as a safe haven is maintained by ensuring that the air pressure in the Service Tunnel is kept above the pressure in the Running Tunnels so that clean breathable air is always present. In addition, when a Cross Passage Door is opened for evacuation in the event of fire, there is a flow of air from the Service Tunnel into the Running Tunnels. This creates a 'bubble effect' clearing smoke from around the Cross Passage Door allowing passengers safe exit from the train.

When a train is evacuated, passengers leaving the train step on to a Tunnel walkway along which they proceed to the nearest Cross Passage Door to enter the Service Tunnel. They are then evacuated from the Service Tunnel into the opposite Running Tunnel via the opposing Cross Passage Door where they will board an evacuation train. This is usually a Passenger Shuttle in normal service which has been requested by radio to stop at the appropriate point.

The two Running Tunnels are also interconnected by 2 metre diameter piston relief ducts at 250 metre intervals, their purpose being to relieve a build-up in air pressure from the piston effect of moving trains. The air is pushed through the ducts into the opposite tunnel reducing air resistance and improving passenger comfort. The piston relief ducts are fitted with dampers which can be closed in an emergency to prevent smoke being pushed from one Running Tunnel to the other in the event of fire. Crossovers about 1 kilometre into the Tunnel at the English and French ends also interconnect the tracks in the Running Tunnels to allow single track working. If maintenance is required on the Running Tunnel North then trains from England to France can be sent via the Running Tunnel South for a period, after which normal working can be re-established from France to England. This alternate working can continue until the maintenance is complete. It is normally carried out during the night.

All traffic passing through the fixed link is controlled by the Rail Control Centre (RCC) at the Folkestone Terminal with a duplicate standby control centre at the

Coquelles Terminal in France. The Folkestone Centre is operated by controllers working in shifts using the rail traffic management (RTM) system for rail traffic and the engineering management system (EMS) to control the fixed infrastructure equipment. All staff in the Control Centre are bi-lingual in French and English.

Four types of train operate in the Tunnel:

1. HGV Shuttles carrying lorries and other goods vehicles loaded on to semi-open wagons.
2. Tourist Shuttles carrying passengers in cars and coaches in totally enclosed wagons.
3. Passenger Shuttles operating between London and Paris and London and Brussels.
4. Freight trains.

It was a HGV Shuttle which was involved in the fire in November 1996. HGVs are driven on to carrier wagons of semi-open construction, each being about 20 metre long and 5.6 metre high. The maximum weight of HGV that can be carried is 44 tonnes. Included in the HGV Shuttle is an air-conditioned Amenity Coach, located immediately behind the leading locomotive, which is used to carry HGV drivers and two train crew members. Total carrying capacity of the Amenity Coach is 52 passengers. The crew comprises a steward who serves meals and the Chef de Train, who is an operational train manager and is situated at a workstation in the Amenity Coach. The HGV Shuttle driver is in contact with the RCC via concession radio and can also contact the Chef de Train by an emergency telephone. The Chef de Train also has access to a public address system in the Amenity Coach.

The tunnels are provided with two ventilation systems:

1. A normal ventilation system which is used to ensure good air quality in the Running Tunnels and to pressurize the Service Tunnel with clean air.
2. A supplementary ventilation system which is intended for emergency use with powerful fans which are capable of being configured to force air through the tunnels in either direction. This enables smoke from a fire on a stationary train to be kept clear of sections of the Tunnel occupied by other trains.

Fire detection systems are provided in the Running Tunnels comprising ultraviolet and infrared flame detectors, smoke detectors and carbon monoxide sensors. The alarms are transmitted to the Fire Management Centre and the RCC. There are two levels of alarm, unconfirmed and confirmed. The activation of a single detector operates an unconfirmed alarm in the Fire Management Centre. This is investigated before evacuation and other emergency measures are implemented, in case of a false alarm. If more than one flame detector alarm operates then a confirmed alarm is also sounded in the RCC.

13.3.3 The accident

13.3.3.1 *Initiation and alarms*
At about 21.30 hours on 18 November 1996, 29 HGVs were loaded onto an HGV Shuttle train at the French terminal at Calais. The train left at 21.42 hours entering the Running Tunnel South at a speed of 57 kilometres per hour. As the train entered the

Tunnel two security guards at the portal noticed a fire beneath one of the HGVs and reported this to the French Control Centre. This Centre then immediately informed the RCC at Folkestone. At the same time an unconfirmed in-tunnel fire alarm sounded in the Fire Management Centre followed by four more in-tunnel fire detection alarms at intervals along the track. The RCC informed the HGV Shuttle train driver that a possible fire had been detected on his train. Almost immediately a fire alarm sounded within the driver's cab indicating that there was a fire in the rear locomotive. This was confirmed by an alarm at the Chef de Train's workstation in the Amenity Coach. By this time the train was travelling at 140 kilometres per hour.

At 21.52 hours another train, comprising a single locomotive, also entered the Running Tunnel South behind the incident train. Within a few minutes of entering the Tunnel, the driver ran into dense smoke causing him to slow down. He was instructed by the RCC to make a controlled stop ensuring that he was adjacent to a Cross Passage Door. A Tourist Shuttle Train followed the single locomotive into the Tunnel. At this point the incident train was still travelling towards the English end of the Tunnel at a speed of 140 kilometres per hour.

13.3.3.2 *Emergency response*

As the RCC had now received a confirmed fire alarm, the emergency response procedures were put into operation and a message was transmitted to all trains to reduce speed to 100 kilometres per hour in accordance with instructions. This was followed by an order to close all the piston relief duct dampers between the two Running Tunnels. In addition, no further trains were allowed to enter either of the Running Tunnels. The RCC also remotely activated the closure of the crossover doors at the English and French ends of the Tunnel. These doors, which should have been left closed, were in fact open due to operational problems.

At 21.56 hours the driver of the incident train noticed an alarm indicating that a circuit breaker at the rear locomotive had opened. This resulted in the stopping of the incident train at 21.58 hours, by chance, close to a Cross Passage Door. However, due to smoke enveloping the front locomotive, the driver was unable to see the number on the door, and was therefore unable to inform the RCC of his precise position in the Tunnel so that the door mechanism could be released. The driver of the incident train informed the RCC that he had lost power. He attempted to leave the locomotive cab to identify the Cross Passage Door but was driven back by dense smoke. The Chef de Train also attempted to leave the Amenity Coach by the rear door in order to locate the Cross Passage Door so that an evacuation could be organized, but was also driven back by smoke. This allowed smoke to enter the Amenity Coach affecting the occupants.

As a result of the fire on the now stationary incident train, the tunnel telephone network between England and France failed. From this point on all communications were restricted to the concession radio which very quickly became overloaded. The driver of the Tourist Shuttle Train, which was the last train to enter the Running Tunnel South, was ordered by the RCC to proceed to the rear locomotive and reverse out of the Tunnel. This was to allow the supplementary ventilation system to be started and configured to force air through the Tunnel towards the incident train driving the smoke which enveloped it out of the French portal of the Tunnel. However on reaching the rear locomotive, the driver of the Tourist Shuttle reported that he had lost traction power. The reason was that the fire on the incident train had burnt through the catenary (or overhead cable) as a result of the train being brought to a stop. Traction power was eventually re-established to the catenary in the area of the Tourist Shuttle and the train was able to reverse out of the Tunnel to the French terminal. The RCC opened the Cross Passage Door in the vicinity of the single locomotive in the Running Tunnel South and ordered the driver to evacuate to the Service Tunnel. By this time there were only three trains left in the Tunnel, the incident train, the single locomotive and a Tourist Shuttle in the Running Tunnel North which was to be used as an evacuation train.

The supplementary ventilation system was then started up in Running Tunnel South in order to ventilate the Tunnel from the England to France direction. Unfortunately the pitch of the fan blades was set to zero and therefore no air was actually blown into the Tunnel. After a long delay the fans on the supplementary ventilation system were reset and a flow of air began to clear the smoke from around the incident train. It was then possible for the Chef de Train to organize the evacuation of the 26 passengers from the Amenity Coach into the Service Tunnel and from there to the evacuation train waiting in Running Tunnel North.

13.3.3.3 *Intervention of emergency services*

The emergency services, initially from the French emergency centre, arrived at the scene at 21.56 hours followed by an English response team at just after 22.00 hours. When the first rescue personnel arrived at the scene it was found that the incident train driver had not yet evacuated and was still in his cab. He was led to safety by the emergency services at 22.30 hours at which time it was also confirmed that there was no one left on board the incident train. For the next 5 hours the emergency services tackled the fire working in relays under extremely cramped conditions made hazardous by falling debris. The fire was finally extinguished by about 05.00 hours the next day.

13.3.3.4 *Consequences of the fire*

Although no fatalities resulted from this incident, the people who were trapped in the Amenity Coach inhaled smoke and toxic fumes and seven of them required intensive oxygen therapy in the Service Tunnel before they could be taken out by ambulance. The remainder of the passengers and crew left the scene on the Evacuation Train. All were taken to various hospitals in the region of Calais and the more seriously injured taken by helicopter to Lille. The periods spent in hospital were short and all patients had been discharged within a few days. Following the incident all commercial services through the tunnel was suspended while investigations were carried out.

As a result of the fire severe damage was caused to 10 shuttle wagons and HGVs, the HGV loader wagon and a locomotive. In addition the fire severely damaged the tunnel structure and ancillary equipment over a length of 2 kilometres of the tunnel. Limited operations were resumed from 29 November 1996, using single line working in both directions in the Running Tunnel North, while repairs were carried out to the Running Tunnel South. The Channel Tunnel Safety Authority (CTSA) carried out an inquiry into the fire. The CTSA is a safety body which advises the UK and French Intergovernmental Commission on all matters concerning safety in the Channel Tunnel operations. The Inquiry Report was published in May 1977 and the following discussion is based on the findings of this report (Channel Tunnel Safety Authority, 1997).

13.3.4 The causes and development of the fire

The cause of the fire in the HGV on the incident train was never established with any certainty. However it was established that the wagon of origin was probably No. 7 from the front of the rear rake of the train (the train comprising two rakes of carriages separated by a flat loading wagon onto which lorries are embarked via a ramp). However, the No. 8 HGV on the rear rake was carrying frozen fat and as the fire grew on the No. 7 vehicle, the movement of the train cause the flames to spread backwards engulfing other HGVs, including No. 8 which produced an extremely high fire load and significant quantities of smoke. As long as the train was in motion, the fires on the HGVs were supplied with large quantities of moving air.

As the incident train began to slow down in response to instructions from the RCC, smoke started to move forward and engulfed the forward part of the train. By the time the train stopped at the cross passage marker the smoke was so dense that the driver could not see the walkway along the side of the Tunnel only 1.5 metres from his door. Once the train had stopped, the flames caused damage to cabling and piping along the walls and roof of the Tunnel eventually causing the overhead catenary to fail with the loss of traction power. There was considerable fire damage to the structure of the tunnel wall.

The smoke from the fire continued to move forward a further 6 kilometres towards the English portal until it was reversed by the operation of the supplementary ventilation system. When the supplementary ventilation fans were started and the blades were configured correctly the smoke was pushed away from the front of the incident train. However, the fire was also fed with large quantities of fresh air and grew in size spreading further down the train. The most significant combustible loads carried on the train were frozen fat and clothing together with combustible materials in the cabs of HGVs and their load of diesel fuel.

Smoke from the fire penetrated the Running Tunnel North via the French Cross Over Doors which were open. According to instructions the Cross Over Doors were supposed to be kept closed, unless one of the Running Tunnels was closed for maintenance and the crossovers were being used. It should have been possible to close these doors remotely but poor reliability in the door mechanisms meant that the Tunnel had been operated with the doors in the open position for some considerable time. In addition, smoke travelled into the Running Tunnel North via a piston relief duct in the vicinity of the incident train because the isolation damper had failed to close. It was later found that out of the 196 piston relief ducts, 19 dampers had failed to close leading to a large number of alarms on the operator screen in the RCC.

13.3.5 The emergency response

Although the accident did not result in any fatalities, it was the first time that the emergency response procedures had been tested by a major incident since the Channel

Tunnel had been opened. It revealed a number of major deficiencies in the response resulting from errors in the way the procedures were implemented. It also revealed a number of problems in the design of the Tunnel Systems and in the HGV Shuttle trains in the event of a major fire. Some of these problems had already been highlighted and had been a matter of concern for the CTSA. Other new and unexpected problems were only revealed as a result of the emergency.

13.3.5.1 *The design of the HGV Shuttle wagons*

It was recognized in the conceptual design that the worst-case scenario for the tunnel was an HGV fire in one of the Running Tunnels. The effects could be catastrophic since at peak operating times up to 10,000 people might be present in each Running Tunnel mainly in Tourist Shuttles and Passenger Trains. The original design of the Channel Tunnel HGV Shuttle wagons specified that HGVs were to be carried inside a totally enclosed steel box fitted with fire detectors which, when activated, would cause halon gas to be injected into the enclosure. This was a condition of the Concession Agreement made between the English and French governments and the operator, Eurotunnel. It had the advantage that a fire would be contained within the enclosure preventing its spread to other vehicles and making it impossible for smoke to enter the Tunnel. In addition, the enclosure could be flooded with halon to extinguish the fire thus preventing damage to the wagon. This would allow the train to be driven out of the Tunnel into special emergency sidings at the exits where fires and other emergencies could be dealt with away from the main terminals. The main objective in the event of a contained fire on a train was a to remove it from the Tunnel as quickly as possible in order to eliminate the potential hazard to other trains.

When the HGV Shuttle wagon design was being finalized, it was realized that if the wagon was to transport 44 tonnnes HGVs, the total weight would exceed the maximum allowable axle loading for the track. The problem was overcome by reducing the weight of the steel enclosure. Instead of using a solid steel box, a lightweight semi-open mesh design was adopted. This decision was made on commercial grounds but had the following consequences for safety:

1. It was no longer possible to extinguish the fire with halon gas due to the lack of a total enclosure.
2. The motion of the train would cause air to fan the fire which would then grow rapidly.
3. The fire would spread to other HGV wagons while the train was in motion.
4. Smoke would enter the Tunnel increasing the risk to passenger trains and Tourist Shuttles.
5. Damage to the shuttle wagon(s) by an on board fire could cause the train to stop in the Tunnel.
6. Fire on a stationary shuttle wagon could result in damage to cabling, catenary and other services running through the Tunnel reducing the effectiveness of emergency response.

The modified design of HGV Shuttle wagon breached a number of design safety principles which were essential to the safe operation of the tunnel. In an attempt to limit the consequences for safety of the modified design, additional fire and smoke detection points were installed in the tunnels, as well as air sampling devices and infrared detectors placed on some of the shuttle wagons. It was also argued that other rail tunnels in Europe did not carry HGVs in closed compartments, but on flat wagons. While the extra fire and smoke detection might have provided earlier warning of a fire, it did little to compensate for the dangerous effects of a fire once it had developed.

Although HGVs and freight trains carrying 'dangerous goods' are banned from the Tunnel, these goods mainly comprise inflammable hydrocarbons and nuclear or toxic materials. The carriage of combustible items such as furniture, textiles and animal products including frozen fat, all capable of producing large quantities of toxic smoke and fumes in the event of fire, is not restricted in any way nor do special precautions have to be taken. The incidence of fires on HGVs was expected to be very small and it was something of a surprise when this fire occurred within a few years of the Tunnel opening. It was perhaps unfortunate that a load of frozen fat was being carried by the HGV adjacent to the HGV that was the source of the fire. However, if separation could have been maintained between the source of the fire and the fuel for the fire (the frozen fat) by using steel enclosures as originally envisaged, it is probable that the incident train could have been driven from the Tunnel without a full emergency response being triggered. Fires on HGVs can be expected to occur from time to time. It was not so much the fact that a fire occurred but the modified design of the HGV Shuttle enclosure that led to the disastrous consequences in terms of injury to passengers, damage to the Tunnel infrastructure, loss of revenue, loss of public confidence and risk to passengers on other trains.

The triggering of the emergency response also revealed a number of incident response errors due to the deficiencies in the emergency systems and procedures. These are described below.

13.3.5.2 *Emergency systems and procedures*

The report of the inquiry into the fire (Channel Tunnel Safety Authority, 1997) is critical of the handling of the incident by the RCC, concluding that numerous mistakes were made. However the report acknowledges that the Control Centre staff was confronted by a series of fast-moving events which completely overloaded their capacity to take stock of the situation and at the same time undertake 'a rapid series of actions in a short period of time'. In this respect the deficiencies in the way the incident was handled are very similar to those in the cockpit of Swissair flight SR111 described in Section 13.2. This is not unusual in emergency situations. Section 13.2 describes how this can stem from a lack of appreciation by designers and managers of the effects of high workload and stress in emergency situations. The normal expectation that people will carry out their tasks successfully most of the time in routine situations, must be suspended for emergency situations. The following problems were identified by the Inquiry Report.

Procedures

The responses required in the event of an emergency were almost entirely rule based requiring access to written procedures which were difficult to follow, extremely complex and badly laid out. These were mainly set out in the form of flow or logic diagrams which in theory can provide a more accessible overview than written procedures if the procedure is not too complex. However, in the case of the emergency procedures for the Channel Tunnel, extreme complexity led to a significant delay in implementation. When immediate actions are required in an emergency, there will be little or no time available to work through complex flow diagrams or written procedures. Although the events may occur with a very low frequency it is important that the basic responses are to some degree committed to memory avoiding reliance on written materials that are referred to infrequently. This is not say that emergency tasks will be carried out at the skill based level; they are unfamiliar and not sufficiently well practised. However, as defined in Section 2.3.1.2 of Chapter 2, rule based actions do not have to be written down but can to some extent be committed to and recalled from memory enabling them to be carried out more swiftly.

An earlier response to the first fire alarm received could have prevented two more trains entering the Tunnel behind the incident train. The entry of these trains prevented the supplementary ventilation system being started earlier (which would have enveloped the following trains in smoke). The earlier use of the fans would have prevented the front of the incident train being enveloped in smoke and allowed faster and safer evacuation from the amenity coach.

Another method of supporting operator tasks in emergencies is by automating simple tasks to release personnel to deal with the more complex knowledge based task of keeping abreast of a rapidly changing situation and making appropriate responses. For instance, an important knowledge based diagnostic task was to assess the position of the fire and the trains in the affected Running Tunnel to decide the appropriate configuration of the supplementary ventilation system. It might be advantageous to automate the selected configuration using a push button to select the required sequence, rather than have to set it up manually. The operator would then be released to deal with the flow of incoming information and subsequent tasks.

Indications and alarms

Information overload is a common feature of emergency situations. Due to the abnormal situation the instrument panels in the RCC were generating dozens of alarms that, first of all, had to be acknowledged but which the operators then had to prioritize in order to make a timely response. At the same time there were numerous radio and telephone calls arriving at the Centre as a result of the large number of people involved in the incident. The Inquiry Report states that insufficient operators were present in the RCC to handle all this information. It recommends that task analyses be carried out for the emergency operating tasks to confirm that sufficient staff are available to carry out the required tasks within the time limitations.

Fire alarm policy

The delay in triggering the emergency response was due to a policy of not responding immediately to unconfirmed fire alarms. This policy had been adopted as a matter of expediency due to the large number of false alarms which had occurred following the commissioning of the Tunnel. These had led to unnecessary activation of emergency procedures causing disruption of normal operations. The Inquiry Report recommended that in future, the incidence of the first fire alarm must trigger an immediate response.

Operation of trains in the event of fire

The original policy had been to drive the affected train out of the Tunnel in the event of a contained fire. All fires would now be uncontained and the procedure for fire on an HGV Shuttle train required the train to be brought to a controlled stop in order to uncouple the front locomotive and amenity coach from the rest of the train and drive it out of the Tunnel. However within a few seconds of the incident train stopping, the fire had burned through the catenary and the loss of power prevented the locomotive being driven.

The response of the RCC to the trains in the Running Tunnel South was to reduce the speed of the trains but allow them to exit the Tunnel at the English end. The effect of this was to draw smoke from the fire over the amenity coach and front locomotive of the incident train. The response would have been correct had it been possible to uncouple the front locomotive and amenity coach so that they could also exit the Tunnel leaving the fire behind.

This is typical of the problems that can arise when using a rule based procedure to deal with an emergency situation. The rules may be so rigid that they only fully apply when the emergency progresses in the way expected by the rule book. Even a slight divergence from the expected behaviour of trains, fires and people can make some of the rules inapplicable. The failure to stop trains ahead of the incident train is an example of this. It caused the front of the incident train to become enveloped in smoke and could have led to fatalities. Only the timely and fortuitous opening of the correct Cross Passage Door by the RCC ensured that there were injuries rather fatalities. The Inquiry Report recommended that Eurotunnel should abandon the 'drive-through policy' for trains.

Infrastructure failures

A number of serious infrastructure equipment failures added to the problems of handling this emergency. One of these was a failure to ensure that the crossover doors were kept shut during normal running. Due to mechanical problems and the requirement to have maintenance staff present while the doors were being closed, a decision had been taken by default to leave them open. This defeated a primary safety feature of the Tunnel, whereby if a fire occurs in one Running Tunnel, then the other Tunnel can be completely isolated as a smoke free safe haven into which passengers from the affected Tunnel can join an evacuation train. If a significant number of passenger trains became trapped in a Running Tunnel with a fire the large number of people to be evacuated into the opposite Tunnel would make such an arrangement imperative. The integrity of this safe haven was also affected by the failure of a significant proportion of the piston relief dampers to close. Train drivers in the Running Tunnel North during the fire reported passing through light smoke in the region of the open piston relief dampers and through dense smoke in the region of the open Cross Over Doors.

13.3.6 Conclusions

13.3.6.1 *Direct cause*

The direct cause of this incident was not a fire on an HGV, something to be expected, but the way the fire developed as a result of a modification made to the original design concept of the HGV Shuttle wagons. With the original design of shuttle wagon, any fire which occurred could be contained within a steel enclosure where it would be extinguished by injection of halon gas. This justified the Eurotunnel policy of attempting to remove the train from the Tunnel to emergency sidings. However this policy was no longer effective once the semi-open design of the HGV Shuttle wagon was adopted for operational and economic reasons. The semi-open design meant that a small fire could not be extinguished and due to the movement of the train as it was driven from the tunnel the fire would grow as it was fanned by large quantities of air.

Table 13.2 Human error analysis of the Channel Tunnel fire

Cause	Description	Type of error or other failure	Systemic cause or preventive action
Direct cause	Initiating event – fire on a HGV developed into a major conflagration as a result of HGVs being carried in open sided shuttle wagons.	Design	Original design concept of a total enclosed shuttle wagons with halon injection in the event of fire was discarded for operational and economic reasons.
Root cause	Reference had to be made to complex written emergency procedures in the form of flow diagrams which caused serious delays in the response.	Active errors – response	The emergency procedures did not take account of high workload. Essential actions in an emergency should be committed to memory to enable a swift response.
	A large number of indications and alarms, together with radio and phone communications flooded the control centre causing information overload.	Active errors – response	Automate simple tasks wherever possible to release operators to carry out knowledge based activity. Limit and prioritize the number of alarms and indications occurring during emergency situations to reduce information overload
Contributory causes	Due to earlier false alarms, there was a policy of ignoring the first fire alarm received which caused a significant delay in the response.	Latent error – management	Operational and economic factors took priority over safety.
	Failures in maintenance prevented the Running Tunnel North being fully isolated from the source of smoke.	Latent error – maintenance	There was a failure to respond in a timely way to poor reliability of essential safety features (Crossover Doors left open, piston relief duct dampers failed to close).

13.3.6.2 *Root causes*

The root cause of the failure to respond effectively to this accident was the extreme complexity of the emergency procedures and the way they were presented resulting in significant delays in implementing the initial response. This was compounded by a system that overloaded the operators with a large number of alarms, indications and phone/radio communications with no means of prioritization.

13.3.6.3 *Contributory causes*

Among the contributory causes to the failures in response were serious failures of infrastructure equipment including a failure of remote closure of the Crossover Doors, which required the presence of a technician. In addition, a significant number of piston relief duct dampers failed to close when operated remotely. These were essentially latent errors in maintenance that had remained uncorrected for a considerable time. They were faults that had no discernible effect on Tunnel operations until an emergency occurred.

13.3.6.4 *Consequences*

The human consequence of the accident was injuries with fatalities only narrowly averted. In addition there were economic consequences of loss of revenue for a considerable period following the accident. More difficult to quantify is the potential loss of public confidence due to safety concerns. No information is available which could confirm whether such a loss of confidence has taken place. However annual traffic statistics published by Eurotunnel for the years following the accident show that by 2002, the number of cars carried through the tunnel had declined by 28 per cent since 1999 although HGV traffic had increased by 30 per cent and the number of foot passengers (Eurostar) had remained about the same.

13.3.6.5 *Human error analysis*

Table 13.2 tabulates the main human errors in responding to the accident in terms of the direct, root and contributory causes.

References

Channel Tunnel Safety Authority (1997). *Inquiry into the Fire on Heavy Goods Vehicle Shuttle 7539 On 18 November 1996*, London: The Stationery Office.

The Transportation Safety Board of Canada (2003). *Aviation Investigation Report – In-flight Fire Leading to a Collision with Water, McDonnell Douglas MD-11 HB-IWF, Peggy's Cove, Nova Scotia, 1998*, Toronto: TSB Canada. Available: http://www.tsb.gc.ca/en/reports/air/1998/a98h0003/a98h0003.asp.

14

Conclusions

14.1 Human error and blame

Human error is not inevitable but rather is the inevitable consequence of defective systems. These systems may be technological systems such as an aircraft cockpit display, a chemical plant control panel or the design of a system for venting an underground chamber. The systems may also be organizational, whether in the form of an operating or maintenance procedure, a checking system for errors made by others, or indeed a high-level management system or the safety culture of an organization. In all cases, the systems influence the way work is carried out and the probability with which errors occur. As there is always a level of residual or random error in any task (see Section 1.2) this probability can never be reduced to zero. The best possible human machine interface design or the most superlative and fault-free management system will never entirely eliminate these residual errors from human activities. However, careful attention to system design adopting a user-centred approach (see Section 4.1) will minimize the probability of the systemic errors which are superimposed above the base level of random error.

At first sight the whole subject of human error appears to be quite impenetrable and highly unpredictable. As a result many managers surrender to the apparent inevitability of errors and revert to the easier and simpler remedy of allocating blame. Blame places the responsibility for an error with the individual making the error. This removes the need to understand why the error occurred since it is believed future errors can be prevented by punitive measures against the individual. If the error was system induced, as most errors are, then this solution to the problem will always be unsuccessful. The defective system will remain uncorrected and it is only a matter of time before another error is committed by the next unfortunate person exposed to it. This is not to say that there is no place for blame. Clearly there is, but the important point is that the blame is attributed where it is deserved. There will be always be cases where the individual making the error deservedly attracts some blame. There may be an element of carelessness, inattention, negligence or deliberate violation of rules that must be dealt with. However, it is important that this is addressed as a secondary issue subservient to a thorough investigation of the possible systemic causes of the error.

Companies and/or industries which over-emphasize individual blame for human error, at the expense of correcting defective systems, are said to have a 'blame culture'. Such organizations have a number of characteristics in common. They tend to be secretive and lack openness cultivating an atmosphere where errors are swept under the carpet. Management decisions affecting staff tend to be taken without staff consultation and have the appearance of being arbitrary. The importance of people to the success of the organization is not recognized or acknowledged by managers and as a result staff lack motivation. Due to the emphasis on blame when errors are made, staff will try to conceal their errors. They may work in a climate of fear and under high levels of stress. In such organizations, staff turnover is often high resulting in tasks being carried out by inexperienced workers. The factors which characterize a blame culture may in themselves increase the probability of errors being made.

The opposite of a blame culture is an 'open culture' which is prepared to admit to and learn from mistakes. It is a culture open to the possibility of error and it encourages staff to report errors they have made so that the underlying causes can be investigated and corrected. In a blame culture, the possibility of punitive action encourages staff and managers to cover up their errors and discourages investigation. Of course an open culture also means that when the systemic causes of error are revealed remedial measures may become necessary and these can be expensive. It is probable, however, that the consequence of further errors will be far more costly in the long term than correcting the problems. Even in an open culture, employees may still be reluctant to disclose their errors to management. They may believe career prospects could be jeopardized or some other form of discrimination practised. Due to this reluctance, which is an all too human trait, a number of organizations have introduced a confidential reporting system (CRS). A CRS enables an error or other safety issue to be reported confidentially by an employee to a third party. The third party then communicates the information to the employer. All CRSs work on the principle that it is less important to identify the employee who has made a mistake than to find out that a mistake has been made.

In the UK, a confidential reporting system (CHIRP) has been in place for commercial airline pilots and air-traffic controllers since 1982 (CHIRP Charitable Trust, 2002) and complements the Civil Aviation Authority (CAA) Mandatory Occurrence Reporting system. The information received is disclosed to those who have the power to take action to correct the problems but only with the approval of the person reporting the issue and ensures that the reporter cannot be identified. It is then disseminated more widely. Consultation for a CHIRP system in the marine industry commenced in early 2003 (CHIRP Charitable Trust, 2003).

More recently, a CRS for the railway industry has been introduced (CIRAS, 2000). From 1 June 2000 CIRAS became a mandatory national system for all UK railway group companies. It provides an opportunity for staff to report their own or others' errors or safety concerns with assurance that no punitive or disciplinary action will result. CIRAS operates in a complementary way to the normal reporting channels although it transmits a different sort of information, in particular, human error and safety issues.

These systems are not limited to the reporting of human errors but also provide a vehicle for staff to report any serious safety concerns, dangerous practices, situations or incidents that might otherwise not be drawn to the attention of management. A CRS is, in essence, a half way step between a 'blame culture' and a completely 'open culture'. While the latter may be more desirable, it is also more difficult to bring about in practice since it requires a fundamental change in company culture. Such changes require a high level of continuous commitment from board level downwards and may take decades rather than years to achieve. It is interesting to note that the industries where a CRS is currently operating or planned tend to be uniformed pseudo-military style organizations where hierarchical rule based structures exist (i.e. airlines, railways and shipping). Such structures, as already discussed earlier in the book, do not sit easily with the openness that is needed for staff willingly to report their own errors.

14.2 Understanding human error

Understanding human error and its systemic causes is a more satisfactory approach than attributing blame to individuals. It is also a more difficult and daunting approach. One of the major problems in understanding errors is the wide variety of forms and types in which they appear. To help make sense of human error a range of classifications or taxonomies have been developed into which specific errors can be placed. An error usually fits into more than one classification; for instance a slip or error of commission may at the same time be a latent error because the consequences are not immediate. These classifications are able to define the forms of the error. In addition there are error types such as maintenance, management or design errors. It is known that most maintenance errors are latent in form. It is also known that latent errors are difficult to detect if they are not discovered quickly. Classifying errors in this way provides useful information and promotes understanding. In turn this opens the way to identification of the systemic causes of the error and ultimately the remedial measures to reduce its probability.

The case studies in Part II showed how it was possible to undertake a human error analysis of an accident. This is called the 'retrospective approach' investigating the human error causes after an accident in order that future accidents might be prevented. The importance of differentiating between the direct cause and the root cause of the accident was emphasized. Since most of the accidents examined in the case studies were due to human failure, the direct cause or initiating event was usually, but not always, a human error. In some cases, the initiating event was a system failure which demanded a human response. Such cases were specifically considered in Chapter 13, although many other case studies involved human responses once the accident sequence had been initiated. If the direct cause was human error, then it was found in most cases that the root cause was the deficient system which induced the error, or at least made it more likely. The root cause was usually discovered by examining the performance shaping factors (PSFs) influencing the particular activity. The PSFs can be identified by reference to PSF classifications or checklists such as that provided in Chapter 4, Section 4.2.4. The PSFs may of course become immediately obvious from discussions with the person committing the error (where this is possible). Most people intuitively understand the factors that influenced the error that they made. Many people strive continuously to overcome negative factors in the work environment perhaps making frequent errors which they are able to recover from on most occasions thus avoiding an undesirable outcome. However, even error recovery has a probability of failure and eventually the consequence of the original error will be realized.

It is rare that accident causation can be simplified to a single direct cause and root cause. The case studies demonstrated the wide range of influences at work in causing an accident. Quite often, for the accident to occur, these influences needed to come together in a specific way at a particular point in time. This coincidence of events underlines the strong element of chance in accidents. Particular accidents are by nature random and unpredictable. Sometimes it can be seen in advance that the required elements for an accident are in place. The poor condition of the UK rail infrastructure, as indicated by an increasing incidence of broken rails was of grave concern to the regulatory authorities long before the Hatfield rail accident occurred. However, it could not have been predicted that the accident would occur at Hatfield, although in terms of probability, Hatfield would have been high on the list due to the particularly poor condition of the track at that location. A failure of management judgment in delaying track replacement due to the negative implication for train operations, loss of revenue and the ensuing penalty payments was the direct cause. The root cause was the difficulty in managing the complex interfaces between the infrastructure controller and maintenance contractors under the post-privatization arrangements. However, this was further complicated by vague definition of responsibilities and a conflict between profitability and safe operation. While Hatfield was an accident waiting to happen, it could not have been predicted with any certainty that it would occur at Hatfield.

Part II of this book illustrated the 'retrospective approach' to understanding human error and provided a qualitative method of human error analysis of accidents. The case studies drew upon many of the concepts discussed in Part I. Part I also demonstrated the possibility of a 'predictive approach' to understanding human error using quantitative modelling techniques. The reader is referred to more detailed texts for the use of particular methods of quantification and modelling, since this book was never intended to be a manual or primer for these techniques (see References). However, Part I illustrates how comparative estimates of human error probability (HEP) can be useful in assessing the vulnerability to human error of particular operations or activities.

Part I also showed the importance of examining human errors in the context of other errors or tasks taking place at the same time and in the same vicinity. Human errors never occur in isolation. Not only are they influenced by the work environment and associated PSFs but they are made more or less probable by preceding tasks and their outcome. Error dependency was shown to be an important factor in determining probability and a dependency modelling technique was introduced to assess this. The converse of dependency, error recovery, was also discussed. Fortunately, most errors are recoverable, but when the consequence of an error is so severe as to be unacceptable, then error recovery may have to be engineered into the system rather than relying on chance or inventiveness.

The automatic warning system (AWS) used on all UK trains provides an inbuilt system of error recovery. If the danger signal is not acknowledged, then the train is brought to a halt. However, it was shown how even this system is vulnerable to failure through over-familiarity (conditioned response) and dependency (having passed a caution signal, the probability of a signal passed at danger (SPAD) is increased). The four aspect signalling system used in the UK was analysed using the human reliability analysis (HRA) event tree technique described in Part I. HRA event trees are probably the most powerful tool available for analysing a sequence of tasks (such as a procedure) in order to identify the human errors making a significant contribution to the overall failure probability. They are especially useful because they are able to represent in a logical way the interactions between errors including dependency and recovery. Due to the inherent complexity of human error and the variety of ways in which it can occur any method which can diagrammatically represent these interactions in time is useful for increasing understanding. In some respects the result of the HRA event tree in terms of numerical estimates is less important than the process of constructing the tree in terms of promoting understanding.

14.3 Human error in industry

The universal concepts needed to understand human error can be applied to any human activity in any industry. It has been shown how maintenance and management errors are usually latent in nature and if they are not discovered at the point of occurrence then there may be little chance of recovery when the delayed effects occur. The case study of the multiple engine failure on the Royal Flight was a perfect illustration of this (see Chapter 9, Section 9.2). While human errors occur in much the same way across industry, not every industry has arrived at a proper understanding of human error. An industry that has always been in the vanguard of understanding human error has been the nuclear industry. This industry has been instrumental in developing many of the techniques described in Part I, particularly those concerning quantification of human error. The nuclear industry has in general had an excellent safety record not least due to its recognition of the potential for human error in causing reactor accidents. However, when something does go wrong, as at the Three Mile Island nuclear facility in 1979, then the adverse publicity generated can have an effect lasting for decades. Long after Flixborough, Hatfield, Ladbroke Grove and Kegworth are forgotten; Chernobyl, Three Mile Island and the Windscale Fire of 1957 will be remembered. It was the Three Mile Island accident which was the main impetus for most of the research into human error described in Part I of this book.

The same motivation to come to an understanding of the role of human error in accident causation has occurred in other industries. The UK Offshore Oil and Gas industry was brought up short by a sudden, unexpected and traumatic accident in 1988, the Piper Alpha fire and explosion in the North Sea which cost 167 lives. As a result new regulations were imposed upon the industry forcing companies to take better account of human error and its potential for disaster in the close confines of an oil and gas platform. This led to major changes in the way platforms are designed and opened the way for new thinking about how best to evacuate them. The railway industry also came to terms with human error and its implications following the Clapham Junction accident in 1987. In a similar way to the Offshore Oil and Gas industry, railways were forced to examine the role of human error and under new Safety Case regulations they were required to make this aspect of their operations more explicit.

Since the earliest days of flight, the aviation industry recognized the importance of human factors in safety. Human factors topics and the vulnerability of pilots to human error forms an integral part of their training. The self-sufficient lone aviator high in the clouds pitched against hostile elements dependent for survival only on his flying skills still forms the popular image of the pilot. The truth is less glamorous; safe commercial flying today is essentially a team activity and the inadequacy of the 'lone aviator' model has been demonstrated in tragic ways. The use of cockpit resource management (CRM) techniques by airlines emphasizes the importance of teamwork at the expense of individualism.

Conspicuously absent from the case studies in Part II is the Health industry. There was a practical reason for this. It may be indicative of the culture of the industry that most reported accidents, certainly resulting in single patient fatalities, are investigated internally and the reports are not publicly made available. Errors in clinical practice are referred to within the industry as 'adverse events'. These are defined as unintended injuries or fatalities due to medical mismanagement rather than by the disease process. The results of a study in UK hospitals in 2001 showed that 10.8 per cent of patients admitted to hospital experienced an adverse event, around half of which were preventable (Neale et al., 2001a). About 8 per cent of these events resulted in a fatality (Vincent et al., 2001b). When this is spread over the 8.5 million patients admitted to UK hospitals in a year it amounts to almost 75,000 deaths per annum as a result of medical error or mismanagement. It needs to be acknowledged of course that patients are highly vulnerable to mistakes being made in their treatment. Nevertheless, by comparison with most other activities, the chance of death on admission to hospital due to an accident caused by medical error, rather than the disease process, is extremely high. The cost to the National Health Service of these errors is estimated at about £1 billion per annum excluding any costs of litigation and compensation. The types of errors involved included diagnostic errors, accidents in surgery, errors in administration of drugs and postoperative care on the ward. There is clearly room for improvement particularly since the Health industry still largely operates a culture where admission of error depends upon medical negligence being proven in a Court of Law. However, when blame is accepted steps are usually taken to understand and correct the conditions which led to the error. The concern is that, in the cases where litigation does not ensue, then the error will remain unacknowledged and uncorrected.

There are wide variations in how human error is understood and dealt with across industry. To some degree this depends upon the prevailing culture within the industry. However, it also depends upon the degree to which individual managers, designers, safety experts, engineers and other professionals understand the issues discussed in this book and elsewhere. Hopefully, the motivation for this will come from a desire to learn the lessons of history as illustrated by the case studies in Part II. If this does not motivate greater understanding, then pressure will inevitably come from increasingly well-informed and discriminating consumers wielding economic power. If this book results in a better understanding of why human error causes accidents then it will have fulfilled its main purpose.

References

CHIRP Charitable Trust (2002). www.chirp.co.uk/air/default.htm

CHIRP Charitable Trust (2003). www.chirp.co.uk/marine/HTML/General Summary.htm

CIRAS (2000). www.ciras.org.uk

Neale, G., Woloshynowych, M. and Vincent, C. (2001a). Exploring the causes of adverse events in NHS hospital practice. *Journal of the Royal Society of Medicine* **94**, pp. 322–330.

Vincent, C., Neale, G. and Woloshynowych, M. (2001b). Adverse events in British hospitals: preliminary retrospective record review. *British Medical Journal*, Volume 322, 3rd March 2001, pp. 517–519.

Appendix: Train protection systems

A.1 Introduction

This appendix provides background information in support of Chapter 10, Active Errors in Railway Operations. It provides a brief introduction to the systems currently in place to protect against signals passed at danger (SPADs), and some systems which have been proposed.

A.2 Protection against train collisions

A.2.1 The time interval system

From the earliest days of the railways, the greatest hazard to passengers has been a collision between two trains occupying the same section of track. The earliest method of obtaining separation between following trains was the time interval system, whereby a signalman (or in the early days, a railway policeman) would be stationed at the side of the track with flags, lamps and a pocket watch. When a train passed, the time would be noted and following trains would be halted until sufficient time had elapsed such that the forward train would have passed the next person down the line. The signalman would not know of course whether the forward train had actually passed the next person, and because of the unreliability of locomotives at that time, rear-end collision accidents were a common event. The introduction of the electric telegraph and ultimately the semaphore signal, remotely operated from a signal box, eventually overcame the hazards of the time interval system, which was soon replaced by the 'block system' in use today.

A.2.2 The block system

The block system imposes a space (rather than a time) interval between successive trains. A length of track is divided up into blocks or sections where entry into each section is controlled by a signal. This warns following trains if the section ahead is occupied. Traditionally it used a combination of interlocking points, semaphores and block telegraph instruments, operated by a signalman, to prevent two trains being in a section at the same time. Once a train has passed on to a section of track, a signal, originally a semaphore which was raised (danger) or lowered (proceed), is automatically set to danger behind the train preventing any following trains entering until the first train has passed out of the end of the section. When the section is clear, the signal changes to caution, or 'proceed with care'. Active errors by signalmen were a common cause of train collisions for many years until semaphore signals and manually operated signal boxes were eventually replaced with more modern automatic fail-safe systems. These use multi-aspect colour-light signals to indicate the state of the section to train drivers and continuous track circuits to indicate positively whether a train is still in a section. Remote computerized railway control centres co-ordinate and monitor operations.

A.2.3 Multi-aspect signals

Most modern railways in the UK and Europe operate with multi-aspect signals on main lines with moderate or heavy traffic. These colour-light signals are operated automatically by track circuits which detect the presence of a train in a section. When a train enters a section an electrical current running through the track is short circuited by the leading wheel set. This causes the signal at the entry to the block to show a red aspect. If there is no train in the block then the signal shows a green aspect. However, for speeds above about 30 m.p.h. the driver would be unable to stop a train in the time it would take from observing the red signal to entering the next section. For this reason, most signal systems will be three or four aspect. The most common is the four-aspect system comprising red, single yellow, double yellow and green aspects.

The four-aspect signal system, uses a caution signal at the entrance to the section ahead of the section which is occupied by a train and protected with a red signal. This cautionary signal has a single yellow aspect advising the driver to expect a train in the section after the one ahead. The four-aspect signal system also uses an additional section protected by a double yellow aspect signal. The double yellow aspect signal will advise the driver to expect a single yellow signal in the next section and a red signal in the section after that. Thus, if a green signal aspect is seen, the driver knows that there is at least three blocks clear ahead. This allows for much higher train speeds while still providing sufficient distance for trains to be brought to a standstill before reaching a section occupied by another train. It also means that slower trains can run closer together thus increasing track capacity. This is not of course a method of automatic train protection (ATP), but merely a system for providing information to the driver about the occupancy of the track ahead beyond visual sighting distance. On observing a double yellow caution signal, it is still necessary for the driver to begin to reduce speed in case a single yellow protects the following block. This will enable the train to be brought to a standstill in case there is a red signal in the block after that.

A.2.4 The automatic warning system

The train automatic warning system (AWS) was developed following nationalization of the railways in 1948 and first came into operation in 1958. It was progressively installed on most UK trains and track over the next few decades along with improvements to signalling. It remains in operation, virtually unchanged. It comprises a magnetic ramp placed between the rails, some 200 metres ahead of a signal, and a detector on the train. The detector can receive data from a pair of magnets in the ramp. One is a permanent magnet, the second an electro-magnet linked to the signal. If the signal is green the electro-magnet is energized. When the train first passes over the permanent magnet the detector on the train arms a trigger to be ready for a brake application. If the signal is at green, then when the train passes over the energized electro-magnet the trigger is disarmed. A bell rings in the driver's cab and a circular visual display shows a black indication. The driver has no need to apply the brakes.

If the signal is caution (yellow or double yellow) or danger (red), the electro-magnet in the ramp is de-energized and when the train passes over it, a siren sounds in the cab and the disc displays alternate black and yellow segments (resembling a sunflower). The audible alarm and display are intended as a warning to the driver that there is a signal ahead to which a response may be required. The driver must acknowledge that the audible alarm has been heard by pressing a plunger on the control desk within 2 seconds to cancel the warning. If the driver fails to cancel the warning then the brakes of the train are automatically applied, the assumption being that there has been a failure to notice the signal. When the alarm is cancelled the display returns to its normal black configuration so that the driver has a visual indication or reminder that a warning has been cancelled. The AWS is, however, unable to distinguish between cautionary (yellow and double yellow) and danger (red) aspects of the signal.

A.2.5 Train protection and warning system

In 1994 a joint rail industry project for SPAD reduction and mitigation (SPADRAM) identified the train protection and warning system (TPWS) as a possible improvement measure. In 1996 a tender was put out to the supply industry outlining the system that the rail industry would like to see installed. A system was accepted in 1996 and development commenced culminating in the start of a number of trials on Thameslink trains with 20 signals on their routes.

The system comprises two induction loops set in the track near the signal. These transmit a low frequency radio pulse when there is a red signal ahead. When the train passes over the first or 'arming' loop, a receiver on the train detects the pulse and a timer is started. When the train passes over the second or 'trigger' loop, the timer is stopped.

The distance between the loops is set so that a train running at a safe speed will take 1 second to pass over both loops. Hence, if the timer stops in less than 1 second then the train is travelling too fast. This causes a full emergency brake application to be made for a minimum period of 1 minute. The driver can only reset the system after this period and after seeking guidance from the signaller. The train will therefore be stopped. The technical name for this type of system is a 'speed trap'.

The system is highly effective at speeds up to about 45 m.p.h. Up to this speed, the train will be stopped in the overlap beyond the signal (usually 200 yards). The overlap is equivalent to the safety distance where there is no danger of collision. If a train approaches a red signal at more than 45 m.p.h., or the safety margin is shorter than typical, then an overspeed sensor is also provided on the approach to the signal. This utilizes an additional arming loop after the trigger loop. This additional loop similarly transmits a low frequency radio signal to the train if the signal is red. The train is able to use these radio signals to determine if it is approaching the red signal at a high speed and again will apply the brakes before the signal is actually passed. As before, once the brakes are applied the driver is unable to release them for 1 minute. Setting up the system is extremely simple. It requires the setting of a timer on the train, depending upon the rolling stock type and setting the distance between speed trap loops on the track.

Up to speeds of about 75 m.p.h. TPWS will still stop a train within the safe prescribed overlap. Over this speed TPWS will still apply the brakes, although the train may not be stopped within the safety distance. However, even at speeds in excess of 75 m.p.h., the system will significantly mitigate the effects of a collision by slowing the train. In fact it is possible for TPWS to work at speeds as high as 125 m.p.h., but with very limited effectiveness.

A.2.6 Automatic train protection

ATP is a system which automatically protects trains against exceeding speed limits and virtually guarantees that all trains will be stopped before a red signal. It comprises equipment on the track which transmits target speeds to trains, and a display in the cab which shows target and actual speed and checks that the train is not exceeding them. The target speed is the safe entry speed to a section of track. In conventional train operation, train drivers receive information and driving instructions from the external environment by observing permanently installed trackside signals. With ATP colour-light signals are no longer required. The driver is presented with a continuous display showing maximum allowable speed for the section ahead and receives advance information about the section speeds beyond that. Rather than having visually to sight a signal outside the cab, all the information will be presented on the in-cab display. Outside the cab, the driver will only see a trackside marker. In Europe for instance, this would comprise a blue sign with a yellow chevron indicating the end of the block where the train would have to be brought to a halt.

For trains running at speeds much in excess of 125 m.p.h., as occurs on some lines in continental Europe, it is not feasible to use conventional colour-light signals visually sighted from the cab. At these speeds, by the time the driver observes the signal it is too late to make the necessary speed reduction or stop the train unless excessively long blocks and overlaps are used, which would reduce line capacity. The same system components can be used to facilitate automatic train operation (ATO), in theory allowing automatic control of the complete journey between stations with the possibility of driverless trains. Using ATO, the role of the driver, assuming one is used, is reduced to that of a minder who simply monitors train operation. Such systems using driverless trains are already in use in the UK, including, for instance, the Docklands Light Railway in London.

ATP capability includes:

- maintaining safe distances between trains,
- applying emergency braking in the event of a safe speed being exceeded,
- monitoring train speed through temporary or permanent speed restrictions,
- monitoring direction of travel to prevent rolling in reverse,
- monitoring of train stop within the destination area of the station and release of doors on reaching the correct position.

The disadvantages of ATP are that it is very expensive to install and may prove impossible to install on trains not originally designed for it. ATP has in the past been disliked by railway managers since with older systems, such as those used on trial in the UK, it can reduce line capacity by as much as 25 per cent. However, more modern systems such as those used in Europe, may in future allow an increase in capacity dependent on the speed of working. Drivers may resent the loss of responsibility implied by a system that delivers instructions about the speed at which they should be driving and overrides their actions if they exceed it.

Index